PRAISE FOR *Silvohorticulture*

'Dibben and Raskin offer an insightful, inspiring but also exceptionally practical guide to integrating trees into annual cropping systems. They demonstrate how stacking a farm upwards has benefits beyond just multiple levels of harvest. From soil moisture management to shade, windbreaks and more, this book will inspire gardeners at all scales to consider what adding trees to the operation could do for them. More importantly, Dibben and Raskin will teach you exactly how to do it. It is always incredible to me how much food can be grown on a small plot of land, and in this book we see how the phrase "the sky's the limit" has never been more applicable.'

JESSE FROST, author of *The Living Soil Handbook*

'Every now and then a book comes along that adds a whole new dimension to a subject that you thought you understood. This book shows us the need for farming and growing in three dimensions; it tells us not only the myriad benefits of interplanting with trees and shrubs but also how to set about doing so in clear and extremely helpful detail.'

JOHN CHERRY, cofounder of the
Groundswell Regenerative Agriculture Festival

'Here are all the reasons why integrating trees with vegetables is a complete no-brainer. The only downside to reading *Silvohorticulture* is that you'll wish you'd started sooner. You knew it was a good idea all along. Here's the proof. Start today.'

JOHN PAWSEY, writer and organic farmer,
Shimpling Park Farms

'*Silvohorticulture* will be part of a resilient farming future in these unpredictable times of climate change. Dibben and Raskin give us detailed explanations about how to integrate trees into our vegetable and fruit farms to add more diversity, ecological function and spatial dimension. With its many practical examples, tables and a great tree directory, this book makes it easy to take the next step into more ecological farming methods.'

HELEN ATTHOWE, author of *The Ecological Farm*

'In this immensely practical guide, the authors provide the technical details towards integrating trees with vegetables and other crops, all while dispelling any preconceived notions that this idea reduces productivity or function. In fact, these systems might prove essential for farm resilience in a changing climate.'

STEVE GABRIEL, author of *Silvopasture*

'An essential read for anyone serious about establishing a truly food-secure, resilient and abundant farming system for the future. Using their combined decades of experience, Andy and Ben demonstrate in detail how trees can and must become a foundational and integral part of our farmscape. The need to plan, and to plant, has never been more urgent.'

LYNN CASSELLS, coauthor of *Our Wild Farming Life*

Silvohorticulture

Also by Ben Raskin

Let's Plant and Grow Together:
Your Community Gardening Handbook
(Leaping Hare Press, 2024)

Plant a Tree and Retree the World
(Leaping Hare Press, 2022)

The Woodchip Handbook:
A Complete Guide for Farmers, Gardeners and Landscapers
(Chelsea Green Publishing, 2021)

Zero Waste Gardening
(Frances Lincoln, 2021)

Bees, Bugs and Butterflies:
A Family Guide to our Garden Heroes and Helpers
(Roost, 2018)

Grow: A Family Guide to Growing Fruits and Vegetables
(Leaping Hare Press, 2017)

Compost: A Family Guide to Making Soil from Scraps
(Roost, 2014)

Silvohorticulture

A GROWER'S GUIDE TO Integrating Trees Into Crops

ANDY DIBBEN
BEN RASKIN

FOREWORD BY
IAIN TOLHURST, MBE

Chelsea Green Publishing
White River Junction, Vermont
London, UK

First published in 2025 by Chelsea Green Publishing | PO Box 4529 | White River Junction, VT 05001 | West Wing, Somerset House, Strand | London, WC2R 1LA, UK | www.chelseagreen.com
A Division of Rizzoli International Publications, Inc. | 49 West 27th Street | New York, NY 10001 | www.rizzoliusa.com
Gruppo Mondadori | Via privata Mondadori, 1 | 20054 Segrate (Milano), ITA

Publisher: Charles Miers
Deputy Publisher: Matthew Derr
Project Manager: Natalie Wallace
Developmental Editor: Sally Morgan
Copy Editor: Angela Boyle
Proofreader: Jacqui Lewis
Indexer: Linda Hallinger
Designer: Melissa Jacobson
Page Layout: Abrah Griggs
Illustrator: Kate Verity

ISBN 978-1-915294-36-4 (paperback) | ISBN 978-1-915294-37-1 (ebook)
Library of Congress Control Number: 2024044827 (print)

Instagram: @ChelseaGreenBooks and @RizzoliBooks | Facebook: @ChelseaGreenPub and @RizzoliNewYork | X: @ChelseaGreen and @Rizzoli_Books | YouTube: @ChelseaGreenPub and @RizzoliNY

Printed in the United States of America.
10 9 8 7 6 5 4 3 2 1 25 26 27 28 29

To our friend and mentor Alan Schofield.
Thank you for inspiring and supporting us both
in our horticultural journeys.

CONTENTS

FOREWORD

At last, a comprehensive manual for horticulturists aspiring to integrate trees into their farms – agroforestry at its finest. I feel a little cheated – why was this book not available to me when I took the somewhat headless plunge into planting trees with vegetables? I could have benefited so much if this book had been around then, saving myself precious time, sleepless nights ('which tree species?') and procrastination. I thought about it for a decade before actually planting, but there was nowhere much to go for information about connecting trees and vegetables. So many decisions to make but so little experience to draw upon. This book shows that the art and culture of agroforestry for horticulture is well and truly established.

I have known Andy and Ben for a good length of my farming career, and I am pleased that they have come up with this book, as they have forged a beautiful writing relationship and really produced the perfect work to cover this subject in an exceptionally comprehensive and staggeringly thoughtful way. Two growers with a heap of experience show clearly how this system could work for you on your farm. There is a sound practical element here for the establishment of what will become an enduring component to your farm system.

Taking the plunge into agroforestry has, until recently, been an act of faith, because so little practical experience exists in the world of commercial horticulture. Establishing an element of such longevity as trees to your farm is a serious investment of not only time but also cash and emotions, so the decision needs to be well planned and followed through. Planting the trees is only the start – the rest comes later. Andy and Ben's attention to detail while covering so many potential aspects is heartwarming and hugely encouraging to those new to the concept and very reassuring to those who have already begun the tree journey.

If you have not yet started your agroforestry project, then this is essential reading that will save you much deliberation and uncertainty.

IAIN TOLHURST MBE

PREFACE

Between us we have decades of vegetable- and tree-growing experience on a number of farms of different types. We first met in 2010 when working together at The Community Farm near Bristol. This is a 22-acre farm growing vegetables to supply its own box scheme and local shops. The pair of us quickly realised we had a shared ethos, outlook and sense of humour. Over the intervening years we have remained both personal friends and professional collaborators. We have worked particularly closely on training the next generation of organic growers and more recently on agroforestry. With a mutual passion not just for the craft and science of growing and the natural world but also for communicating that excitement, it was probably inevitable that we would colluded to write a book together.

It has been an integrated writing process, so we hope that you won't be able to spot which bits were written by who. To avoid you having to work out who was talking at any point in the book we have adopted a generic 'we' for all of our shared experiences and advice. Where Abbey Home Farm is mentioned specifically, it is Andy's experience and learning, while Eastbrook Farm is Ben.

We have also both been lucky enough to visit and learn from a wide range of super knowledgeable growers and agroforestry pioneers. We have showcased a few of these examples as case studies, but there are others we did not have the space to feature who are referenced in the acknowledgements. We are aware that there are other numerous wonderful agroforesters across the globe, however our experience is from the UK and so most of the examples and species recommendations flow from that. The principles are universal, so you should be able to build on the content and adapt to suit your climate and local trees and vegetables.

ACKNOWLEDGEMENTS

ANDY DIBBEN

Although they will be unable to ever read this book, I would like to thank my parents Penny and Bob Dibben for instilling a sense of wonder towards the world around me from a very young age. This has resulted in a lifelong thirst for knowledge.

My beautiful wife Sarah and my great lads Connor and Archie have been ever-present sources of support, new knowledge, fun-filled distraction, and inspiration to always try new things, as well as being incredibly tolerant of my constant musings on agroforestry.

My passion in natural history and, in particular, the tiny, wonderful things in this world was fired by John Arnott, whom I was fortunate enough to study under for a year. My old friend Mike Hicks introduced me to the wonderful world of plants, then the inspiration and teaching of both Iain Tolhurst (Tolly) and Ben Raskin converted that passion for plants into a career as an organic grower; both were also instrumental in my introduction to agroforestry.

This book would not be the same without the unique and vivid illustrations drawn by my great friend Kate Verity, beautifully demonstrating the ability of creative arts to inform and educate.

I would especially like to thank Hilary and Will Chester-Master for sharing their space and wisdom with my family and me. It is here at their beautiful organic farm that I have truly realised the power and importance of sustainable food production to heal people and the planet.

BEN RASKIN

There are so many people who have fired my enthusiasm for agroforestry and helped to shape my thinking and approach. The nature of acknowledgements, however, means we have to do the hard job and single out a few for individual mention. Early on in my horticultural career in the late 1990s, I was privileged to work with Hugh and Laura Ermen on their 'own root' fruit trials at Garden Organic's Yalding

Gardens. At the same time Katie Butler, the head gardener there, taught me the fundamentals of pruning and tree growing. This foundation convinced me that all growing spaces should have some trees, and so I have tried to integrate them wherever I have worked since. However, it was not until I started at the Soil Association that I became fully aware of the concept of agroforestry. Like most of us in the UK, the late Martin Wolfe was instrumental in my journey, and a visit to his pioneering agroforestry farm, Wakelyns, in the East of England opened my eyes to the opportunities for larger-scale agroforestry systems. Working with the team at Organic Research Centre, and especially Sally Westaway and Jo Smith (both of whom are now running their own agroforestry systems), gave me the confidence that the science backed up my grower's instinct that mixing it all up with trees made sense.

There are a number of other growers who have helped to build my knowledge and confidence in integrating trees into vegetable cropping. Alan Schofield (to whom this book is dedicated) operated a longstanding and complex agroforestry system in North West England long before the rest of us had heard of the concept. Sadly, Alan's personal circumstances meant we were not able to feature his system in the book in as much detail as we would have liked, but his spirit is very much embedded in it. Tolly (Iain Tolhurst of Tolhurst Organic) is a constant inspiration, still trying new ideas and pushing the boundaries of the possible. Other growers like Martyn Bragg, Marina O'Connell and Matthew Wilson have also been generous with sharing their experience and expertise.

Thanks as always to Helen Browning for enabling our experiments at Eastbrook Farm and to my ever loving and supportive family: Ruth, Ivan and Jonah.

Ancient Technique for Modern Times

'The End is Nigh' is an understandable human reaction when momentous and troubling events are taking place. We are currently in a period of unquestionable and perhaps unparalleled environmental uncertainty. Records for hottest and coldest, wettest and driest weather are being set on an almost weekly basis. At the time of writing, we are emerging in the UK from the wettest winter in 40 years. Many autumn-sown crops have failed or will not reach their potential, while the window for establishing spring crops has been minuscule. We spend time and money cultivating soil, sowing and planting crops. How can we ensure that they survive and provide something to feed people?

For some the enormity of our current situation leads either to denial or to resignation and a feeling of helplessness. For us, as life's eternal optimists, it has the opposite effect: we seek solutions. For us the overarching solution is not in the latest technological fix but in the resilience of the system. Whatever the potential of your high-performing F1 or GM variety, or how much fertiliser and pesticides you throw at it, if the ground is waterlogged or baked dry the crop will not grow. Reductionist approaches to nature are inevitably doomed. We can spend years perfecting a plant that is resistant to the latest problem pest, but, unless it can also grow in wet, dry, and windy conditions, we might find those years of research have been wasted.

Diversity is the key! We should aim for diversity of crops, including annuals, biennials, perennials, but also diversity of genetics within crops, with different varieties and ideally with open pollinated varieties that are able to adapt to a wider range of challenges. The logical extension of this is to bring more woody species back into farming. Trees and shrubs do not need annual cultivation to establish and bring a stabilising effect to growing systems. Trees mostly have a positive effect on the wider environment, for instance through reducing risk of flooding, sequestering carbon or enhancing biodiversity. When thoughtfully integrated, trees will also bring a raft of benefits to you as a grower and to your garden, smallholding or farm. We had previously viewed integrating trees as a useful addition to our growing system, helping to boost yields and plant health. We now see it as the *only* way we can continue to produce consistent yields within this new climate paradigm. Welcome to Silvohorticultural Agroforestry – the future of vegetable growing.

The basic definition of agroforestry is, for us, the 'deliberate combining of agricultural crops with trees'. The term was first used in the mid-1970s in a report by Canada's International Development Research Centre. However, trees and vegetables have played integral roles in human food production ever since we transitioned from gathering food crops to growing our own crops.

The fig tree (*Ficus carica*) is believed by some to be the first plant we domesticated, in the Jordan valley at least 11,200 years ago, pre-dating the domestication of cereals by a thousand years.[1] On the other side of the world in the Southwest Amazon, the guava was domesticated around 9,500 years ago.[2] Meanwhile in Peru and Mexico, there is evidence of the domestication of the first vegetable crops, such as squashes, at around the same time.

In fact, all across the world, as the climate rapidly changed after the last ice age, cultures independent of each other began to domesticate plants and animals and to develop systems for producing their own food rather than gathering wild food.

At broadly the same time as we started domesticating crops, humans developed a production system called 'slash and burn', also known as swidden or shifting cultivation, in which areas of rainforest, temperate forest or woodland are felled and burned to create clearings for crop production. When fertility begins to drop in the clearing, another clearing is created, and the exhausted clearing is allowed to regenerate back to full tree cover. This process is carried out on long cycles across a wooded area. This technique is so effective at producing food that it is still in use today, on more than 280 million hectares (692 million acres) of farmland worldwide, mainly in South America, Sub Saharan Africa and South East Asia.[3] Though this form of food production seems brutal, it has stood the test of time and might provide an alternative view of farming with trees that contrasts to our westernised, modern annual cultivation that is currently destroying soils worldwide.

Agroforestry systems have been used ever since farming began and are still used all over the world. However, agroforestry in general has been in steep decline across the globe over the last 100 years. Increased reliance on larger machinery and chemical inputs since the 1960s has favoured simple monocropping systems over the more complex, dynamic agroforestry approach. However, all is not lost. Agroforestry is having a dramatic resurgence, with modern farmers looking back to old techniques and to traditional farmers who are still using agroforestry for inspiration and guidance. Modern science and cheap flexible fencing options are supporting this move back to more-resilient, nature-friendly approaches.

This historical bank of knowledge is a valuable resource of inspiration for anyone looking to design their own agroforestry system. Although there is a wave of interest in the subject, there are still precious few examples of agroforestry being reintegrated into modern farmland, especially ones focused on intercropping vegetables and trees. We found this a challenge when designing our own systems, and it was the biggest motivation behind writing this book. We hope to draw together information we have learnt on our separate journeys discovering agroforestry. Knowledge is drawn from academic research, real world examples and personal experience. The objective of the book is to take the reader through the whole process of designing an agroforestry system based around vegetable production, sometimes referred to as 'silvohorticulture'. This is perhaps a good moment to outline the different forms of agroforestry that are commonly referred to.

Silvopasture refers to the integration of trees and livestock. Trees bring many benefits to livestock production, providing shelter from wind and rain, and shade from sun, and also offering alternative and highly nutritious sources of feed. The trees also introduce the possibility of adding new revenue streams for farmers or reducing on-farm costs by producing your own fencing and building materials to sell or use.

Silvoarable refers to the integration of trees and arable crops, often in an alley cropping system, where trees are positioned in rows with arable crops between them. In modern systems, the spacing between trees is matched to the mechanical requirements for arable crop establishment and harvest. Trees provide shelter from weather and help recycle and add nutrients to the soil, and again the trees also offer the opportunity to introduce new crop lines.

Silvohorticulture refers to the integration of trees with fruit or vegetable crops. Trees offer many potential benefits to vegetable production: recycling nutrients and adding them to the soil, providing a source of composting materials, boosting pollinator and predator insect populations, managing drainage, preventing soil erosion, reducing irrigation requirements, providing shade and expanding edible crop range. In fact, there are so many benefits you could write a whole book on it!

Some growers and farmers will combine these different systems and, in reality, there is much overlap. However, the latter will be our sole focus for this book. Of the three examples above, it is the integration of trees and vegetables that is the most dynamic system of the three, perhaps offering the most potential benefits and, also presenting the most potential challenges. These will be explored in great depth throughout the book. As we take you through the design process, it will become clear that it is the balance of benefits and challenges that is the real skill of integrating trees and vegetables intelligently.

We have aimed for the content of this book to be relevant to all those who would like to integrate vegetables and trees in the same growing area. We believe the concepts discussed in this book are scalable, as large or as small as you desire; from one tree in a veg garden to thousands across a whole farm, the same basic principles still apply. Our experience has mostly been in commercial production, though, so this is the lens through which we inevitably see.

CHAPTER ONE

Principles and Benefits of Silvohorticulture

Many of the basic principles of agroforestry apply across all the different disciplines, such as managing water, providing shade, reducing soil erosion, enhancing biodiversity, adding new crop lines and mitigating and adapting to climate change. From here on we will focus specifically on the integration of trees with vegetable crops.

Increasing Crop Lines and Yields (without Increasing Land Area)

Perhaps the simplest way to understand how agroforestry can increase crop range and overall yields on the same land area is to visualise the branches of a tree laden with fruit overhanging a bed of vegetables beneath it. This ability to grow crops in the same space at different vertical levels is a powerful image. Vertical stacking of crops is a concept already used by vegetable growers through intercropping, for instance when underplanting tomatoes with basil or growing swede beneath kale plants. It is likely, therefore, that vegetable growers already understand and make use of three-dimensional space. Maybe we have a head start on our arable and livestock colleagues when it comes to understanding and exploiting the benefits of stacking crops vertically. However, to maximise the benefits of integrating trees into our vegetable system, we need to bring trees and woody perennials into the mix.

The most important influential factors in any vegetable production system are light, water and soil. If managed correctly, these factors completely negate the need for any chemical inputs, a benefit that is too often overlooked in modern intensive agriculture. Often however, in a standard vegetable production system, none of these resources are used to their full potential.

Light

Originally, the atoms of carbon from which we're made were floating in the air, part of a carbon dioxide molecule. The only way to recruit these carbon atoms for the molecules necessary to support life – the carbohydrates, amino acids, proteins, and lipids – is by means of photosynthesis. Using sunlight as a catalyst, the green cells of plants combine carbon atoms taken from the air with water and elements drawn from the soil to form the simple organic compounds that stand at the base of every food chain. It is more than a figure of speech to say that plants create life out of thin air.

MICHAEL POLLAN, *The Omnivore's Dilemma*

Light is the most influential factor on plant growth. The process of photosynthesis is the fundamental driver of life on Earth. Without light, there are no plants, and without plants no food or oxygen for animals. For us as vegetable growers, light has a significant effect on both crop yield and quality. How can we maximise the benefits of this completely free and powerful resource?

With a ground-based vegetable system, much of the light falling on the ground is not used by the crop, especially when crops are young and large spaces are left for plants to grow into as they mature. Perhaps surprisingly, many crops don't even need direct sunlight all day for all of their growing life. Lettuce, chard, spinach, cauliflower, broccoli, radish and celery all perform better without direct sun all day, especially during the long hot days of midsummer.

Integrating trees into a vegetable cropping area can make more efficient use of available light. There are two ways trees do this. Firstly, they absorb a proportion of the sunlight as it travels to the ground. If managed correctly, this allows enough light for both tree and vegetable crops beneath them. The second way of sharing light resources between tree and vegetable crops is by using the movement of the sun across the sky through the day and the transient shade it casts to provide enough sun to crops to allow maximum performance. By actively trying to share the sun's rays between leaves at different vertical levels, it is possible to increase yield per square metre, from the same amount of sunlight.

Water

Water is so fundamental to life that it is the first thing we look for on other planets to indicate the possibility of life. Even in the most arid areas of Earth, with little or no rainfall, plants have developed ways of absorbing water from the air.[1] Managing water in cropping areas is one of the most important skills a vegetable grower needs to master. Every

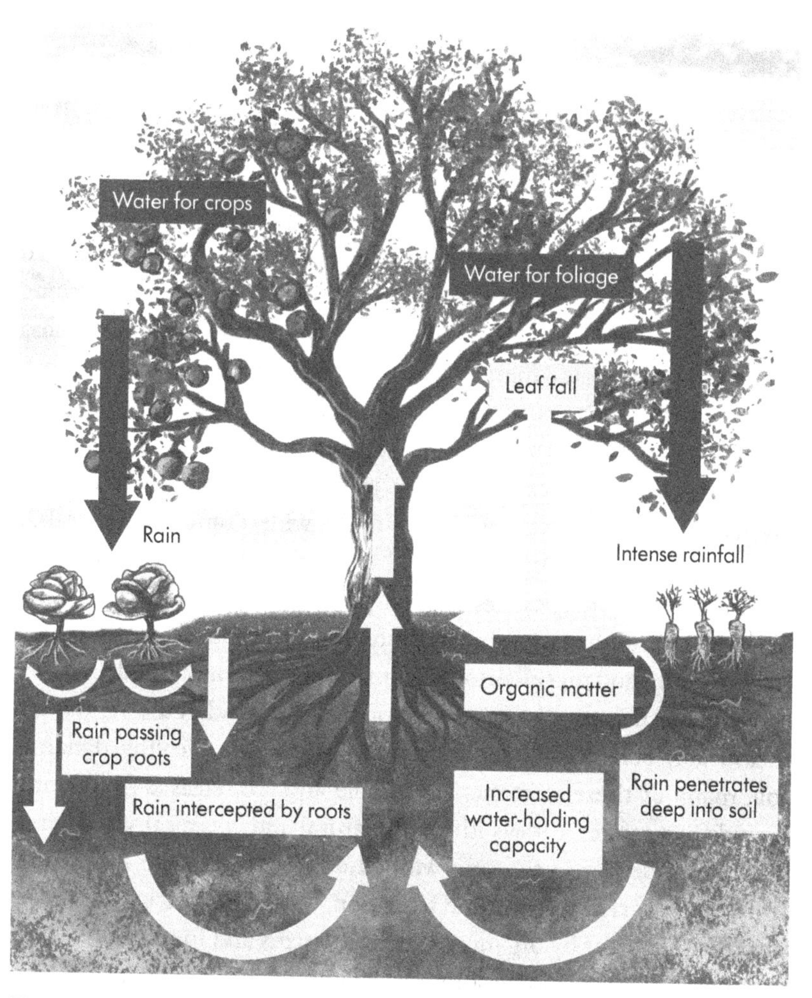

Figure 1.1. Maximising use of water.

grower dreams of owning a machine that controls the weather, but rainfall almost never delivers precisely the amount of water required for the crops at precisely the right time. Usually far more water falls from the sky than our crops can take up. Most annual vegetable crops are relatively shallow rooted, using just the top 30–45cm (12–18 in) of the soil profile.[2]

This minimal depth means a lot of precious water quickly passes down through the soil and beyond the reach of our vegetable crops. If managed correctly the roots of trees can act as a safety net below, absorbing and making use of excess rainfall, ensuring a higher percentage of rainfall is translated into crop growth.

Soil

Soil, unlike light and water, is not essential for crop production. This is clearly demonstrated by techniques such as hydroponics and aquaponics. However, as committed organic producers, for us the soil is essential for truly sustainable crop production. It is one of the most wondrous, mind-blowing and fascinating environments on Earth. If managed correctly soil is a complex, dynamic, effective and self-replenishing medium to grow any crop in. Soilless systems mostly rely on a limited range of human-made mineral, fossil fuel or animal inputs, typically a maximum of 30 or so nutrients. Since humans have been shown to need 130+ nutrients to be truly healthy, we strongly believe that we should all be eating most of our diet from plants and animals produced on diverse, healthy soils that have access to the widest range of minerals.

There are a number of essential mineral elements provided by a healthy soil for plant growth: nitrogen, phosphorus, potassium, calcium, magnesium, sulphur, boron, chlorine, iron, manganese, zinc, copper, molybdenum and nickel. Another four elements found in soil – silicon, sodium, cobalt and selenium – although not considered essential are measurably beneficial to plant growth. In an active, healthy, dynamic soil, many of these elements are mobile and are released from compounds in a range of ways: through natural and chemical weathering at different depths in the soil; washed down through the soil profile from the surface by rainfall and irrigation; or transported up and down through soil layers by organisms such as worms and fungi.

The relatively shallow rooting behaviour of many annual vegetable crops means that many of these crucial elements may not be within

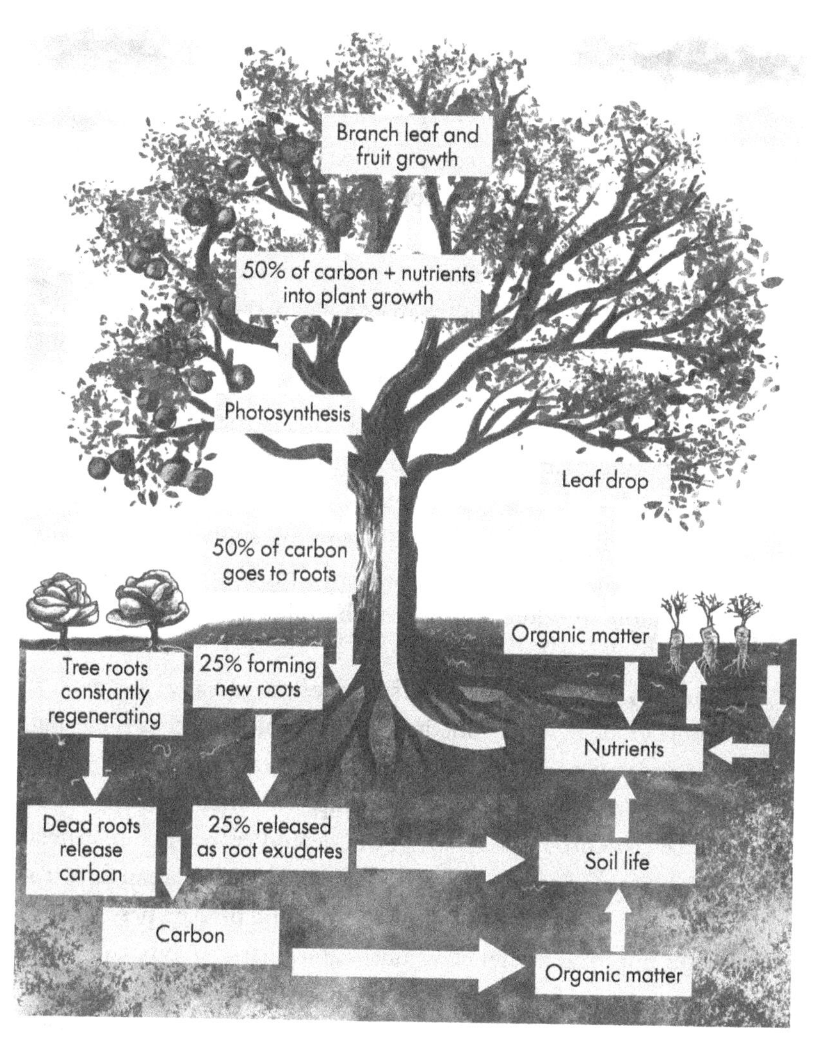

Figure 1.2. How a tree affects soil health.

reach of crop roots or quickly wash past them dissolved in water. As discussed above, the tree roots reach out into cropping areas beneath vegetables, where they intercept and make use of soluble nutrients that escape the clutches of vegetable crops. These same tree roots can access elements being mineralised at depths beyond the reach of vegetable crops. If managed intelligently this interaction between tree and vegetable can lead to increased yields per square metre and reduce or remove the need for imported natural or artificial nutrients.

Enhancing Biodiversity with Trees to Improve Crops

Trees sit at the centre of highly complex food webs and are in themselves multifaceted, diverse habitats. They provide food, shelter and a home for a wide array of biodiversity, including fungi, moss, lichens, bacteria, invertebrates and mammals. These creatures live in, on, above and underneath trees, and for many species trees are an essential part of their existence.

Biodiversity is in catastrophic decline all over the world, with habitat loss being perhaps the biggest singular cause. Sadly, food production has been and still is the major cause of habitat destruction; farming has intensified and increasingly excluded most life from its fields except the intended cash crop. It is logical that if the decline is to be reversed, food producers must evolve their systems to support biodiversity. Trees, especially in the form of agroforestry, provide the opportunity to ensure that food production areas can be valuable habitats as well as being able to feed people.

Sustainable vegetable producers are often already aware that encouraging biodiversity is not just a charitable exercise. Thriving, complex ecosystems teeming with life offer many opportunities for a grower to develop a symbiotic relationship with their system, which will enhance crop production. Fungi can defend crops and are crucial in the mobilisation of nutrients in cropping areas. Bacteria can fix nitrogen into the soil. Invertebrates pollinate crops and predate pest species. Birds are an important defence against caterpillars, rabbits and mice. Mammals can be useful predators of agricultural pests, too. Foxes eat rabbits and catch pigeons; bats are often an unseen ally against many nocturnal agricultural pests.

Agroforestry offers vegetable growers an intelligent way of harnessing and maximising this powerful tool.

Using Trees to Manage Water

We looked earlier at how tree roots help make the most of valuable water resources. However, as well as maximising the water use of cash crops, trees can help regulate and stabilise the water cycle in periods of both excess and dearth, benefiting both the farm and wider society.

Floods

Rainfall is becoming more intense and erratic with climate change. This brings a raft of problems across the world, such as damage or destruction of crops, soil, homes, highways and civil infrastructure. Similar to reversing the biodiversity decline, farmers find themselves in a position to play an effective role in mitigating flood risk. We can directly manage flooding on farms and play a role in slowing the flow of water downstream to alleviate flooding off farm. Planting more trees is a fantastic tool for doing this. The sheer intensity of modern rain events is overwhelming existing drainage systems that have previously been able to cope, whether that be in fields, on a landscape scale around rivers and in urban environments, or even in your back garden. Now we often get longer periods without rain. This causes the topsoil levels to dry out. When heavy rain falls onto that dried soil, the water cannot infiltrate, so run-off and flooding are further exacerbated. Trees can slow and alleviate the effects of heavy and constant rain in a number of ways. During the COVID-19 pandemic, the pressure on hospitals to cope with new cases was as much the problem as the disease itself. Likewise, being able to slow down flows of water during extreme events is so important to avoiding catastrophe.

So how do trees reduce water flow? Firstly, they capture water before it even hits the ground. One only has to take a walk in a woodland on a rainy day to appreciate this. When a tree is in leaf, it will of course intercept more rain, so summer growth and evergreens have greater capacity than deciduous trees in winter. But even a bare tree will hold some water on its branches.

The second way in which the trees slow down the passage of water is once the raindrops have found their way through the canopy and hit the ground. If this ground is vibrantly healthy aerated soil, the rain will be steadily absorbed by the soil. The roots of trees physically keep the soil fractured and open so it can act as a natural soakaway. Beyond these simple physical effects, the roots of trees are a major driver of soil health due to the highly complex web of life that they support in the soil. Through effects such as root exudates and symbiotic relationships with fungi, the positive effects of trees on soil reach far beyond the tips of their roots. All soils, however healthy, will experience saturation events with modern rain patterns, but the length of time taken to reach saturation will be hugely enhanced by the presence of trees and their roots. Holding all this extra water in the soil slows the water cycle, alleviating flooding of fields, roads and housing.

If your ground does reach the point where it is saturated, water will pool and flow across the landscape horizontally. This is the third way trees can slow down the movement of water. Placed intelligently, trees can act as a landscape feature that physically intercepts water flowing on the surface, again slowing its passage to drains or rivers. This effect is described in more detail shortly with soil erosion.

Drought

The compelling argument for agroforestry is, in part, the amazing ability trees have to solve completely opposing environmental challenges. We have just seen how they are a great tool in very wet weather. Conversely, they can be just as useful in extremely dry conditions. The healthy soil around tree roots has the magical quality of being able to drain well but also retain water in all the air gaps present in healthy soil. There, it can be accessed by crops during dry weather and, perhaps just as importantly, be accessed by other soil life that is equally dependent on access to moisture. In comparison, a compacted soil with poor structure and no air gaps will struggle to absorb and hold moisture, whether that water is supplied by irrigation or rain.

The shade and windbreak effects created by trees can also greatly reduce the transpiration and evaporation of water from leaves and soil by lowering temperatures and increasing humidity. These changes conserve precious water during droughts and reduce the need for irrigation. It is a common misconception that sun is the main driver for loss of moisture from soil. In fact, wind is just as much to blame. During high-pressure weather periods in the winter, when the intensity of the sun is low, strong dry continental winds can dry out our soils at a phenomenal rate. When you combine strong winds and strong summer sun, both of which are intensifying equally with climate change, drought conditions and their effects can be greatly accelerated and amplified. Strong winds in cropping areas can also be problematic for carrying out accurate irrigation activities, particularly with sprinkler and rain gun systems.

Reducing Wind Speeds in Cropping Areas

Our vegetable cropping system at Abbey Home Farm is sited right next to where the owners erected wind turbines 25 years ago. Great for energy production, not so good for growing vegetables. Reducing wind speeds was the initial motivation for looking at agroforestry.

High winds cause problems. We've spent many hours chasing mesh crop covers around fields during gales, and wind can inflict serious damage on structures such as polytunnels and glasshouses. But strong winds can have profound negative effects on the crops themselves, too, physically damaging mature plants by either completely flattening tall crops, such as sweetcorn or sunflowers, or toppling trellised and supported crops, such as climbing beans, cucumbers and tomatoes. Small transplants, such as squashes and lettuce, can take longer to establish and grow in very windy conditions.

Plants do not have to be totally blown over to be permanently damaged. Shredded leaves affect visual quality, sale price and saleable yields. Snapped and damaged stems and leaves not only slow the growth of the plant but also create potential entry points for moulds. The constant rocking back and forth of crops in even moderate winds can leave plants unstable and allow their roots to become exposed and damaged. Horticultural and speciality cropping systems are likely to benefit particularly from windbreaks and shelter due to the high values of their output crops. Increased saleable yield ranges from 5 to 50 per cent once wind speed is reduced.[3]

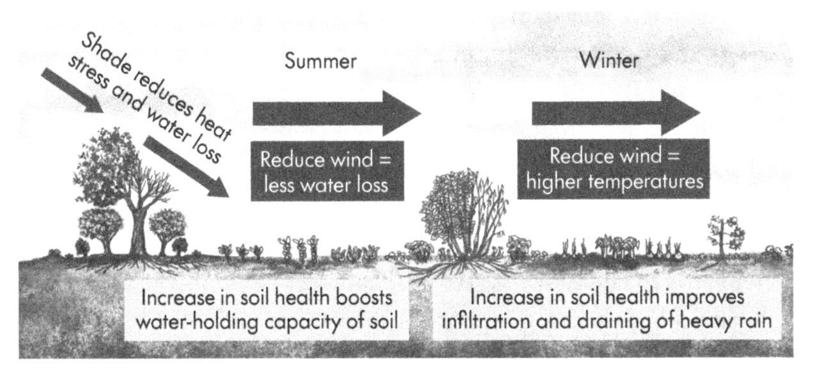

Figure 1.3. How a tree affects shade, wind, flood and drought.

On a biological level, plants grow slower and have smaller leaves in high wind environments. They close the stomata on their leaves to reduce transpiration and loss of valuable water. However, closing their stomata for long periods affects their ability to photosynthesise and respire properly. Plants also experience lower rates of pollination when windy because the tiny pollen particles are quickly whisked away far beyond their intended female recipients. Even for those plants that are insect pollinated, gales can desiccate and damage flowers, meaning they are shorter lived, while pollinating insects are also less effective in gusty conditions.[4]

Creating windbreaks with trees can have noticeable beneficial effects on temperatures in cropping areas by creating pockets of microclimates within the farm or garden.[5] When wind is reduced, soil temperatures are higher, typically by 1 or 2 degrees Celsius.[6] Obviously, this effect would occur downwind of a windbreak; however, although less pronounced, there are also positive effects on soil temperature upwind, showing windbreaks have beneficial effects in both directions.

Providing Shade for Crops in Intense Summer Heat

We looked at how trees can maximise effective use of available sunlight through the vertical stacking of crops. Trees can also provide an important protective role by casting shade. The principle of using trees to shade crops is not a new one. For more than 2,600 years, paulownia trees have been used in China to shade arable crops from summer heat.[7] Intense sunlight can damage crops directly by scalding fruits such as peppers and raspberries. It can also significantly raise temperatures in cropping areas, as discussed in reference to wind speeds.

While raising temperatures can often be desirable, there is a limit to the benefits of increasing temperatures. In most annual vegetable crops, photosynthesis rapidly decreases if temperatures go higher than 34°C (93°F). Once they exceed 37°C (99°F), a plant's ability to regulate water starts to fail. At 47°C (117°F), plant cells begin to die. Dry plants will have a higher leaf temperature than those with an ample supply of moisture. If a plant's water regulation fails, dangerous spikes in leaf temperature can happen in the high 30°Cs.[8]

These are temperatures that all vegetable growers now have to deal with on a regular basis, and hot spells are likely to increase in frequency, length and temperature. Even if temperatures do not reach dangerous levels, growers will need to increase irrigation, which puts extra pressure on water supply and staff time. In fact, during droughts, irrigation often becomes the difference between success or failure, commercially. With water supplies coming under increased pressure worldwide, reducing irrigation requirements is a major focus for many vegetable producers.

It takes time for trees to grow to the size where they provide meaningful shade, so planting trees now to provide increased natural shade should be a consideration for all growers.

Preventing Soil Erosion by Wind and Rain

Each year, an estimated 24 billion tonnes (26.5 billion tons) of fertile soil are lost worldwide due to erosion.[9] That's 3.4 tonnes (3.7 tons) lost every year for each person on the planet. These soils are a farmer's most valuable resource and, if lost from the land, are incredibly damaging to aquatic ecosystems. Soil will silt up river beds, preventing fish from spawning and contributing to algal blooms that disrupt the delicate balance of life in our rivers and oceans. As custodians of the land, we should be doing all we can to keep these soils in our fields.

When a raindrop hits bare ground, it can throw soil particles 0.6m (2 ft) high and 1.5m (4.9 ft) horizontally. This is the beginning of soil erosion and is called 'splash' erosion. As water begins to run across a surface, it collects soil particles as it goes, and this is called 'sheet' erosion. Splash erosion can be addressed and minimised using trees. The effects of the canopy slowing the rate of rainfall on the ground were discussed earlier, and the same process lessens the impact of each raindrop. Improving soil structure and increasing soil organic matter will also reduce this splash erosion. Sheet erosion can be managed very effectively by intelligently positioning tree rows in fields; even though they may allow excess water to pass them, the rows slow and filter the water as it passes through the understorey of trees, keeping soil particles in the field.

Extreme rain is not the only risk for soil erosion, though. Long periods of dry conditions increase the likelihood that we will see our precious topsoil blow away as dust. Perhaps most dramatically demonstrated by the Dust Bowl of the 1930s in the United States, this is a real problem for those growing on lighter soils and those in systems that have been intensively cropped for many years. Using trees to reduce wind speeds and increase organic matter in cropping areas can significantly reduce wind erosion of topsoil.

Agroforestry is not a silver bullet for soil erosion, but it is a powerful tool when employed alongside techniques such as minimising tillage and cover cropping.

Holistic Systems Approach

As experienced sustainable vegetable growers, we have always adopted a 'holistic systems approach' to food production, but what exactly does this mean? It is literally the opposite of conventional or industrial agriculture, which take a more reductionist approach. Rather than isolating a challenge, such as a particular pest or a certain nutrient, and dealing with that on its own, the holistic approach involves looking at the whole growing area as one living organism, whether it be a garden or allotment, a small market garden, one large field or a whole farm with many fields, hedgerows and woodlands. The longer we have had the privilege of working our land, the more we have become aware of the never-ending complex interactions between everything with which we share this planet. Once you embrace this reality, you become aware of how all the different parts of your farm are interacting with each other: animal, vegetable or mineral.

The real skill of a sustainable food producer is to manage all the different, complex interactions happening all over the farm. You must balance the demands of making money, feeding people, employing people ethically, providing habitat for biodiversity, minimising imported inputs. And you must act as a positive force in the landscape for managing water, reducing soil erosion, playing a part in maintaining community and social cohesion, being carbon negative, and the list goes on.

The dynamics are ever changing and require constant modification of the system. The word 'system' is a crucial one here. It would be almost impossible to attend to all of the above challenges separately but simultaneously, yet that is the approach modern agriculture has taken for more than 50 years. With endless consultants specialising in one very narrow subject – nutrition, pesticides, accountancy, machinery, irrigation, employment agencies – the cumulative effect results in the farmer being pulled in many conflicting directions at once.

By adopting a holistic systems approach, many of the subjects can be addressed at once with one approach. Take cover crops. If used intelligently, they can manage soil nutrients, soil organic matter and soil erosion, sequester carbon, provide habitat for biodiversity, and increase rain infiltration. Trees in the form of agroforestry are perhaps the most potent tool for use in a holistic food production system, as has been clearly displayed in this chapter, where so many potential benefits have been revealed. However, designing a good system that allows all of these benefits to be realised demands care and attention. Trees offer so many powerful potential benefits to vegetable production, but, as is often the case in life, this large potential, if not implemented and managed correctly, can cause problems rather than solutions. This is why designing a good system is vital. We hope that this book will provide you with the tools you need to do just that.

CHAPTER TWO

Site and Soil: Assessments and Implications

Every situation where vegetable production and trees are integrated is unique. This challenge was one of the key motivations for writing this book. Off-the-peg solutions won't work. Every location requires a bespoke design, and, when creating this design, we must understand how it interacts with the wider natural landscape. No farm, homestead, field or garden exists in isolation from its surroundings.

In every design, there will be variable aspects that can be altered to a certain degree, such as tree or vegetable choice and the tree understorey. However, there will also be fixed characteristics that cannot be altered, the obvious examples being altitude, climate and soil type. It is these features that we must ascertain at the start of any design process, for they inform layout, tree choice, tree protection, potential production challenges and routes to market. When taking on a new site, we should try to find out as much as we can before planting trees. Soil surveys, meteorological records and even historical maps are all useful. Even better is to talk to previous owners / land managers, who will have an intimate knowledge of the site. Soon after starting at Eastbrook Farm, and having only seen the farm in spring, we were sharing plans for a diverse fruit orchard in one field with Henry, the farm manager who had worked the land for decades. 'Bit wet I'd say', were pretty much his only words, but words enough for me to quickly shift the fruit plan to a different field and change to a more robust mix for the wetter land. His sage advice was the difference between success and failure.

Climate

Climate is perhaps the most influential factor for crop choice and even methods of production, whether for vegetable crops or trees. Although climatic conditions are often talked about in global terms, as growers we know there can be huge localised variation even within short distances. Learn as much as you can about the local weather! The key information is rainfall amounts and patterns, last and first frost dates, highest and lowest yearly temperatures, wind direction and level of exposure to wind on site. Having a good idea of local weather patterns will help inform effective system design.

Vegetable growing is a short-term venture, so we are mainly concerned with current climatic conditions. As the climate shifts, we can react on an annual basis and change crops or variety to evolve our farms. However, the trees will be around for decades, so understanding how climate change will affect your area is also crucial in helping you choose the right trees, species and genetics for the woody part of your plan. Some trees that are perfectly suited now might not grow as temperatures warm, while other species that currently struggle might thrive as the climate changes. In the UK, Forest Research have an online Ecological Site Classification tool that models different climate scenarios and indicates which trees will be suited to your soil and location.[1]

Altitude and Aspect

Altitude impacts the natural tree cover or lack of it in any area. It can also define the variety of species that are growing in an area. Species behave differently at different altitudes. Trees grown at higher altitudes are often slower growing and more compact at maturity. Light intensity increases with altitude due to the thinner atmosphere, so trees that grow at higher altitudes naturally often have silver or shiny leaves to reflect light.

Aspect (or direction) helps define temperatures, too, determining the amount of sun that falls on plants each day and the exposure of the land to prevailing winds. Aspect also affects frost dates. A north-facing slope will be prone to heavy late frosts and lower temperatures. A south-facing slope by contrast will be prone to faster defrosting when

Figure 2.1. Varied farm landscape.

it does frost, which can be just as damaging; if a blossom is frosted, it is much better for it to thaw slowly. Cold air moves down slopes, so where you are situated on a hillside should also be considered.

On a site with existing vegetable production, the current crop orientation needs to be considered. This setup need not be retained in any new design, since in many situations you will be able to change the cultivation direction to suit a new agroforestry system. However, whether keeping your original layout or modifying to fit with tree planting, the orientation of vegetable cropping is a key consideration when planning tree spacing, intended height and shade cast by trees. When designing our system at Abbey Home Farm, trees were integrated into a field that was already in vegetable production. The existing cropping was orientated in an east-to-west direction. One of the first decisions in the design process was to reorientate all cropping to a north-to-south direction, so that when tree lines were integrated we could allow for an equal distribution of shade to nearby vegetable crops.

Topography

The shape of terrain on the site defines agricultural behaviour and therefore the design process. Growing on anything but the most gentle of slopes requires some compromise, especially with mechanised growing systems. Cultivating up and down a slope is the safest and most effective; however, it increases the risk of soil erosion during heavy rain. Many agroforestry systems try to use trees planted across the slopes (ideally even following the field's natural contours) to help collect water in dry periods and slow run-off in the wet times. However, using cultivation or harvest machinery across a slope is often less effective and potentially very dangerous. We'll look at how to tackle this challenge in Chapter 7, discussing how trees can be planted to enable more sustainable soil management on difficult growing slopes.

Natural Capital

In addition to soil and geology, your site is likely to already have a wealth of natural resources growing and living on it. Even environmentally degraded farms have resident plants and animals. This is usually a positive but could also present a challenge. For instance, are there some persistent soil diseases, invasive plant species such as Japanese knotweed (*Reynoutria japonica*), or large populations of deer? Whether the biology of the site is something you would like to support and harness, or conversely to eliminate or manage, you should understand what you have and how it will affect your agroforestry plans.

We will go into much more detail on tree choice in the next chapter, but a crucial part of the design process is to observe what trees grow and thrive locally. We want our trees to grow quickly and well. If there is already a good population of a particular species nearby, we can be fairly sure they will succeed. If you have time, you may even want to collect seeds from local species and grow them on for your own site. It is important, though, to make allowances for future changes in the climate – what grows well locally now may struggle in the future. Knowledge of how different tree species are predicted to react to changes in climate is vital.

Assessing how your site sits in the wider landscape will often give you an idea of potential pests that may affect establishment and successful cropping of trees. If located near large areas of woodland or forest, there is a high likelihood of squirrel and deer populations, both of which can severely impact establishment of young trees and reduce yields on mature trees. Although the presence of pests should not stop you planting trees, it will affect protection and possibly tree choice. It is also important to note if there is any existing fencing or hedging on site, what kind and in what condition, as this may also help protect against pests. A well-maintained, thick, thorny hedge can be as effective at keeping out deer as a fence.

Access

The available access to your site impacts design decisions in a number of ways. Firstly, can you get to the site with a vehicle in all weather conditions? Will the vehicle need to be a 4×4 or tractor? Tree-planting season tends to occur when the ground conditions are most challenging. You will need to get trees, fencing materials, and potentially woodchip for mulching, on and around the site. How wide are the gateways and tracks? Are they surfaced or muddy? One farm we worked with could get to their field only over a narrow bridge. They had to get their woodchip delivered to a nearby farmyard and then ferry the chip across in smaller trailers pulled by quad bikes. This hugely increased the cost of mulching and the time taken.

Beyond the practicalities of planting trees, we also need to design our systems to allow future access for our vegetable production kit. Remember that trees grow sideways, as well as up. When planting next to tracks and gates, leave enough space for the final size of the tree or you will either spend the rest of your years pruning or resigning yourself to scratched tractors and reduced visibility. Think carefully about access in and out of the site, and even within the field. With long rows of trees, it can be worth leaving a gap halfway down to allow access without having to travel all the way down to the end of the field.

Water

If you are adding trees into an existing vegetable-growing enterprise, you are likely to already have some water on site. While not essential for establishing and growing trees, it can be helpful to be able to irrigate young trees during establishment in very dry climates or during unusually dry years. Mostly, we aim not to irrigate, as we hope trees will quickly establish a deep rooting system to make them resilient. Watering can encourage the roots to stay near the soil surface where it is moist. However, in 2018 we planted around 15,000 trees at Eastbrook Farm. In the UK, we had a spring drought that lasted around three months. We lost just over half of the trees we had planted, and the ones that survived were either planted early, mulched heavily or irrigated. The peak time you might need to irrigate newly planted trees is spring and early summer; coinciding with high water demand for horticultural crops, this timing could present problems.

We would recommend that you calculate your current water supply, pressure, source and storage capacity, including the seasonal availability. Also, you may be planning some tree crops that need ongoing irrigation, for instance, smaller fruit trees and bushes, so this, too, will need to be planned for. Our intention in agroforestry settings is to plant trees, including fruiting and nutting species, that are able to stand on their own feet when mature, so mostly we plan only for the possibility of establishment watering. If you are intending to raise trees from seed for planting or sale, then water is essential.

Economic Location

The geographical location of the site in relation to intended routes to markets is another influential factor when designing a commercial agroforestry system. As with vegetables, tree crops have a product shelf life and a distance to the intended customer or processor that will define infrastructure for storage and methods/cost of delivery. As well as affecting profit margins, it may even determine the feasibility of certain crops. For example, producing soft fruit for direct sale if you are a long distance from your customer is tricky, as you would need chilled storage and transport, which adds significantly to your costs. As with any new enterprise, understanding where your markets are, and whether you can produce at a cost that will make them succeed, is helpful. When planting trees, however, it is not always so easy, as you have to wait longer till they crop, by which time markets and prices may have changed.

Soil

Soil can be categorised according to the dominating size of particles within the soil, for example, sand, clay, silt, peat, chalk and loam. It is not uncommon for soil types to vary within one field. Understanding the soil type and the inherent properties of your soil type will inform tree choice and allow you to take advantage of particular soil types or mitigate any associated potential challenges.

Table 2.1. Soil Characteristics

Soil type	Fertility	pH	Water Retention	Potential Challenges
Sand	low	tends to be acidic	Free draining with limited water holding capacity.	Soil erosion, water retention, maintaining fertility.
Clay	high	varies	Slow to drain but prone to cracking when dry. Sweet spot for ground preparation can be fleeting.	Challenging physically to work, slow to warm in spring, prone to structural damage.
Silt	high	varies from slightly acidic to slightly alkaline	Free draining with good moisture retention.	Prone to compaction.
Peat	No one should grow on peat. 1.2 per cent of all UK carbon emissions are from cultivation of peat soils, according to Lowland Agricultural Peat Task Force Chair's report in 2022. It is imperative we leave peat where it is and don't touch it.			
Chalk	variable	alkaline	Varies from light to heavy.	Dries quickly in hot weather, difficult to work when wet.
Loam	high	varies	Good drainage and moisture retention.	Can sometimes be very stony.

Soil pH

Soil pH can be altered through amendments, but this effect is only slight and temporary. As a rule, most trees will grow happily within a pH range of 5.5 to 7.5. Above this range there is a tendency for nutrients – such as iron, manganese, phosphorus and magnesium, even if present in soil – to become inaccessible to plants. Below this range the availability of plant nutrients, such as phosphorus and molybdenum, decreases, and the availability of some elements is increased to toxic levels, particularly aluminium and manganese. Soils below a pH of 4 have a very detrimental effect on beneficial soil organisms, such as nitrogen-fixing bacteria and earthworms.

Although the majority of trees can cope with most soils, there are certain species that will only thrive in high or low pH soils. For instance, sweet chestnut (*Castanea sativa*) and witch hazel (*Hamamelis* spp) prefer acidic soils. More details on choosing the right tree for your pH can be found in the Tree Appendix.

Soil Depth

One of the selling points of agroforestry systems is that trees can mine deeper layers of the soil, but this power is oversold and comes from a common misunderstanding regarding the rooting depth of trees. We see how trees like oaks send down a large taproot to establish themselves quickly and probably also instinctively assume that such a large structure must have deep roots to stay upright. However, huge deep tap roots are a rarity with most trees. For example, Scots pine (*Pinus sylvestris*) and silver fir (*Abies alba*) typically have 90 per cent of their roots within the top 1m of the soil profile. This has two main implications. Firstly, you can successfully grow trees in relatively shallow soils, though they are likely to be less successful and smaller in soils less than 50cm (20 in) deep. Secondly, if tree roots are not going down, they are going out, since what you see above ground is replicated below ground. In a vegetable agroforestry system, this can obviously have an impact on your crops. We'll look at tree rooting behaviour in more detail in the next chapter.

Soil Health

Assessing your soil health early in the design process enables us to take any necessary remedial action before tree health and establishment suffer. There is plenty of advice on testing soils, so we won't try to replicate that in this book. Suffice to say that digging soil pits is the best way to get a sense of what you have. This can be supplemented with lab testing to get a better idea of nutrient levels. Soils in poor health will not establish trees as quickly or as successfully compared with those in good health.

It is helpful to know the history of land use associated with a particular site, as this can be a good indicator of what soil health to expect. Land that has been continuously cropped with heavy machinery will often have damaged soil structure and nutrient shortages, whereas a field that has been in permanent pasture for a long time will generally have good structure and nutrient levels. One of our objectives when planting trees might be improving soil health, so we are not looking to create the perfect soil before planting trees but rather to remove any major barriers to establishment and early growth, such as compaction or significant nutrient deficiencies. The exception to this might be where we are hoping to establish a highly productive fruit system. In this instance, taking more time to ensure the soil is in optimum condition would be sensible. Since we presume that most readers of this book are also growing vegetables, we are supposing that you are not on the worst of soils and that time spent improving soils for trees will also benefit your vegetable plants.

Laws, Regulations and Taxes

In some countries and in certain circumstances, planting trees is regulated or requires certain conditions to be met. For instance, in the UK some plantings require the landholder to do an Environmental Impact Assessment (EIA). Other potential things to consider before planting are whether your trees will prompt a change of land classification from agricultural to forestry, or whether they will have an impact on the value of the land. The latter may not matter to you but could be relevant if the land is held in a trust or shared by a family. Planting trees on tenanted land can also be challenging; even if both landlord and tenant are in principle happy, you will need a tenancy agreement that reflects the risk and reward of planting trees and the variable value of the trees over the course of the tenancy.

It is not possible to cover all of this in the context of our book, and it could quickly become out of date as regulations and taxation systems vary greatly and change frequently. Be sure to check both local and national implications of any planting

Urban Agroforestry

Although people do raise livestock and grow arable crops in urban locations, it is relatively rare. Vegetable production, however, is a feature of every urban environment on the planet, whether it be a balcony, back garden, allotment, school garden, community garden or even urban wasteland. Where there is a space and a desire, people always find a way to grow a few vegetables. Integrating trees into your growing area, however small, can greatly add to the sense of seclusion. Even one tree can provide significant visual screening. Although most agroforestry will occur in rural areas, there are benefits that are particularly useful in an urban environment. In larger community gardens, all the other benefits already discussed in this book can be useful, and, in the context of the urban environment, trees are more effective and far more beautiful at achieving this than a fence. Trees play an important role in providing shelter and food for biodiversity; this role becomes even more important in urban environments where habitats can be rare and isolated. Creating taller elements in garden havens significantly increases their value to wildlife and makes it easier for wildlife to travel between biodiversity pockets.

The majority of towns and cities have not been designed with intense heat waves in mind. In 2022, a report published by the World Economic Forum revealed that scientists have now discovered that cities and urban areas are warming 29 per cent more quickly than rural areas on average.[2] This increase is due to the heat retention qualities of concrete, asphalt and other urban building materials. With many urban growing projects being based around community engagement, filling this role in an ever-warming world will be greatly helped by providing shade to gather and shelter under. Trees are cheaper than buildings and air conditioning units.

Ever since animals began to form social groups, trees have played an important role in the bringing together of these groups. This is particularly true for humans. For millennia trees have played a key social function in almost all cultures, whether sheltering from rain, heat or cold, collecting food or fuel, or often simply as an important meeting point. Trees still play this role in the modern world, perhaps one of the greatest examples of this being the Cherry Blossom Festival in Japan, where a tree's natural process of flowering triggers a nationwide festival and social coming together. Urban environments can often be lonely places and urban growing is a great way to meet others and find community. Trees can play a powerful role in helping create a place for people to come together. There is even evidence that trees can reduce crime in cities through increased socialisation in those spaces and removing wasteland or deserted areas that become no-go zones between criminal factions.[3]

Protected Cropping

Trees are often dismissed as being too large to grow in glasshouses and polytunnels. However, this can be managed by choosing a tree of the right species, variety and rootstock and by suitably pruning. In particular, trees in protected cropping can provide shade during the summer months. Purpose-built automatic shading systems for polytunnels and glasshouses can be extremely expensive to install or, if you design your own, are time consuming to manage and ineffective. Trees can offer a cheap effective way of reducing excessive heat and overly intense light levels. The dappled shade cast by trees can significantly reduce ambient temperatures and light intensity.

It is worth noting that protected cropping environments are a valuable part of vegetable production, protecting some of the most profitable crops from the elements. The beauty of the three-dimensional aspects of agroforestry can be realised well in this situation, as new crop lines can be added above existing ones, allowing growers in frost risk areas to grow early blooming tree crops, such as apricots, peaches and nectarines. Our case study of Troed y Rhiw on page 92 shows how it might be managed.

Protected cropping, especially in Europe and the northern United States, is often associated with high-value summer crops, such as tomatoes and peppers. However, season extension and winter cropping are just as important an economic use for many growers, not to mention the benefits of fresh salads and greens to the health of local communities through the dark months of winter.

Make sure that any trees grown in protected cropping allow high light levels in the winter. In the winter production of vegetables, light is often a more important currency than heat, due to short day lengths. We would choose deciduous trees for this kind of system. To keep light levels balanced between the needs of vegetable crops and trees, we tend to choose trained systems, such as espalier, cordon or fan. Root pruning is of particular importance in this environment, as is targeted watering, and these are discussed in chapters 7 and 8.

CHAPTER THREE

How Trees Interact with Crops and Each Other

The relationship between trees and vegetable crops in a polyculture system is both complex and dynamic. They will interact with each other in space and time, so, before setting objectives and designing a layout, let's try to understand some of those interactions and envisage how they might change as the trees grow and the system matures. While we will never be able to predict or fully understand all that is happening in nature the design process should aim to maximise positive and minimise negative effects between trees and crops.

Fertility

One of the most compelling arguments made about the benefits of agroforestry systems is the potential increase in productivity that can be achieved by including trees alongside your crops. Understanding how to take full advantage of these benefits, however, is not easy and much of the evidence is still somewhat inconclusive. Planting trees and shrubs further complicates already complex natural systems, making it hard to identify which bit of a system is providing the benefit, or perhaps disadvantage.

The three main claims for how trees can enhance the fertility of soil are by increasing:

- carbon, by increasing soil organic matter
- nitrogen, especially by the inclusion of nitrogen-fixing trees
- a range of other nutrients that trees bring up from lower levels of the soil and deposit in the topsoil for use by crops

Let us examine each of these in turn.

Carbon

There are broadly five ways that trees can help build soil carbon.

1. Leaves. Created by taking carbon from the air through photosynthesis, leaves from the tree drop onto the ground around its base and are incorporeated into the soil, thus increasing the soil organic matter. There is plenty of evidence that this theory is indeed true.[1] In horticultural systems, the only downside might be that, for certain crops, an autumnal dump of leaves could damage them (for example, the weight of accumulated leaves could crush a stalk) or risk creating conditions that encourage disease (such as retaining heat and moisture, which could encourage mould growth).
2. Root exudates. A significant proportion of the nutrients that plants produce is not used by the plant itself but exuded through the roots into the soil. The rhizosphere (or area around the tree roots) becomes more biologically active as soil organisms gather around the roots to feed on these exudates. The theory is that this should build soil carbon. However, even this is not so simple; one study suggests that the biological activity boosted by the exudates being pumped into the soil speeds up the carbon cycling and doesn't necessarily result in more stable long-term carbon stores.[2] It does, though, increase soil health and, therefore, should increase the fertility of the soil and our ability to grow crops successfully in this bio-boosted land.
3. Fungi. One Swedish study showed that up to 70 per cent of soil carbon might come from the fungi in it rather than from the plants.[3] Planting trees in cropping systems provides a habitat, undisturbed and rich in root exudates, in which fungi can thrive. What is less proven is how far the carbon built up immediately under the trees moves out into the cropped areas. This spread no doubt depends on crops grown and rotational factors as well as soil and climate.
4. Tree biomass. The carbon built up in the above ground parts of the tree will not necessarily help to build soil carbon. It will depend on how that woody material is used. If left to grow as a tree, it will simply sit in the trunk and branches. If used for wood fuel, obviously the carbon will be released back into the atmosphere. However, the simplest way to utilise the wood for soil health is

to chip any prunings or coppiced material and then add the chips back into the cropped soils. One recent study showed that adding ramial woodchip (RWC) to soil was 'a promising option for SOC [soil organic carbon] storage in soils. The improvement in chemical and physical soil quality, partly due to an RWC amendment, was linearly related to the increase in SOC content. We found the RWC amendment affected SOC quality, which is assumed to have improved soil structural porosity and, in turn, easily available water storage and aeration'.[4]

5. System benefits. The final bit of the picture when it comes to carbon are the benefits the trees provide to the rest of your growing system. This, of course, gets harder to pin down, but mechanisms such as creating microclimates and increasing soil health could enable your horticultural crops to grow better, quicker and for longer, all of which will also have an impact on soil carbon, particularly during the fertility-building phase of a rotation.

Nitrogen

Nitrogen is often the limiting factor in a successful growing system.[5] Despite being abundant in the air, it is usually highly soluble and notoriously hard to hold onto in agricultural systems. Historically, growers built rotations that increased nitrogen during one part of the rotation using nitrogen-fixing crops, such as clover, which they then exploited during the productive phase for their cash crops. The invention of nitrogen fertiliser bypassed this method by allowing farmers to apply nitrogen directly to the soil in a readily available form. However, we

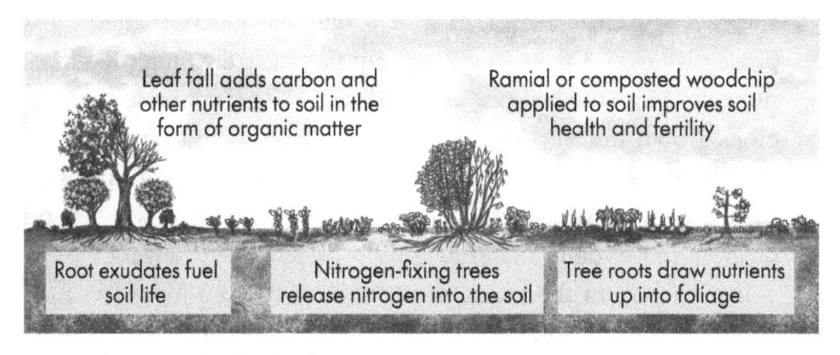

Figure 3.1. Benefits for fertility.

now know that long-term or excessive use of nitrogen fertilisers can cause a range of problems, including nitrogen leaching and pollution,[6] overly rapid growth of plants, making them susceptible to disease,[7] and the depletion of soil health.[8] Can trees offer an alternative source of nitrogen for farmers?

Although it is well known that some trees are nitrogen fixing, growers have historically been nervous about harnessing this ability due to the risk of woody soil additions robbing nitrogen from the soil as they start to decompose. There could be real potential, though, to exploit the nitrogen-fixing abilities of trees to supplement other fertility-building strategies. What isn't clear is how much nitrogen is released by the tree into the system. Soil type and abundance of mycorrhizal communities appear to have an impact on how readily nitrogen moves from a fixing tree into its surrounding environment.[9] One study found that nitrogen-fixing trees can even alter the pH and biology of the soil in which they are growing. Their results indicated that 'the interaction between N_2-fixers and their soil microbial community provides trees within the forest community with improved access to mineral resources that can help meet their high nutrient demands for fast growth and high rates of carbon accumulation'.[10]

Beyond the direct input from tree roots into the soil, which is tricky to predict and measure, there are clear benefits from the nitrogen contribution of fallen leaves. Nitrogen-fixing trees may have 25 per cent higher leaf fall compared with non-leguminous species, and, although only approximately 2 per cent of the leaf is nitrogen, that still represents a useful addition to the system.[11] The biggest opportunity, we believe, comes from using summer prunings from nitrogen-fixing trees. One study found that using the loppings from jumbay (*Leucaena leucocephala*, a leguminous tree estimated to be capable of fixing 500 kg [1,102 lbs] of nitrogen per year) added approximately 80kg/ha/yr (71 lb/acre/yr) of nitrogen to the system.[12] It would appear that using perennial green-manure crops like this could reduce nitrogen losses compared to clover-based fertility systems, but also that the nitrogen is released more slowly.[13] Using fast-growing species, such as alder (*Alnus* spp) or black locust (*Robinia pseudoacacia*), on a rotational coppice could help to build a long-term, self-sustaining fertility regime for your cropping.

Other Nutrients

Unlike carbon and nitrogen, which come primarily from the air, most other nutrients have their source in the soil. Although some soils may have a specific deficiency (for instance, many soils in the UK are low in boron or copper), in general most soils have an ample supply of most minerals needed for good crop growth. The key is making them available to plants at the right time. All the mechanisms we have described above will of course bring other nutrients with them. Leaves and wood are not made up purely of nitrogen and carbon. It is generally stated and accepted (though not experimentally verified) that trees will access nutrients from the lower levels of the soil due to their larger root system, though, as discussed at the end of this chapter, tree roots may not necessarily grow deeper than, say, chicories or vigorous grasses. However, what is clear is that they boost soil health,[14] and specifically fungal populations, therefore improving the nutrient cycling within soil.[15] Much of that nutrient will now be part of the active cycle rather than locked up and will be available to crops.

Competition for Light

All plants use photosynthesis to grow, but too little or too much light can have negative effects on plant performance and, therefore, crop yields. In particular, the shade cast by your trees is a key design factor in vegetable agroforestry systems.

In the UK, day length varies from 7 hours at winter solstice to 16 hours at summer solstice. The angle at which light from the sun falls also changes hugely between these times, as shown in Figure 3.2.

With careful planning, we can design our space to manage where this light will fall and try to eliminate the negative and accentuate the positive effects of that shade. When the tree rows are orientated north to south, all crops will be in direct sunlight for a large proportion of the day as the sun traverses the sky. Conversely, tree rows planted east to west would result in some crops always being in shade and others in full sun all day. Although this might be beneficial for a small number of crops, the consensus has been that north-to-south orientation will maximise light in most cropping situations. One 2018 study explores this in a little more detail.[16] Their modelling suggests that, during

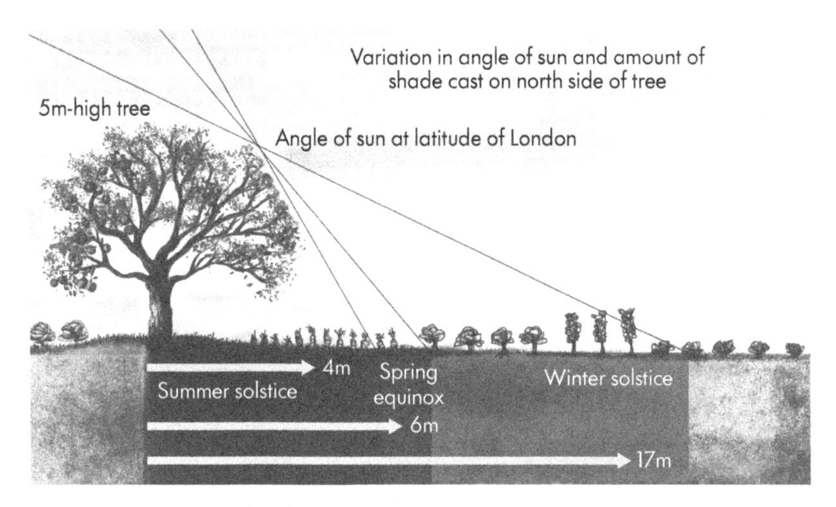

Figure 3.2. Seasonal light angles.

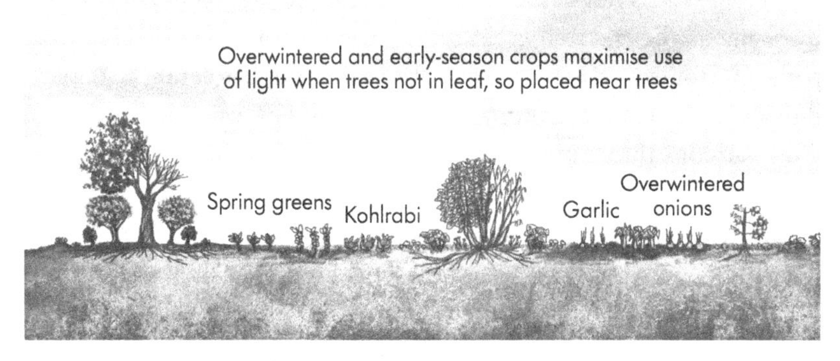

Figure 3.3. Seasonal phasing.

spring and autumn in northern latitudes, a north-to-south planting allows the most light. The trees are in leaf but the sun is still low in the sky, so an east-to-west planting would shade the crops. For the other seasons the orientation is less crucial. In winter, most trees are bare and, therefore, not intercepting much light; while in summer the sun is higher, so the shading effect is reduced. In lower latitudes nearer the equator, where the sun is higher in the sky for longer periods, an east-to-west design is recommended.

The other way of managing the effect of shade from trees on vegetable crops is to plan cropping around the emergence and fall of

Figure 3.4. Different tree habits.

the tree leaves. This could apply particularly to the edge beds under the tree canopy in an alley cropping system. Overwintered crops, such as onions, garlic and spring greens, do most of their growing when a deciduous tree is not in leaf, and will not, therefore, be much affected by shade as the trees come into leaf in the spring. See 'Phasing' on page 122 for more. There are options for shade-tolerant cropping directly under the tree canopy in the tree row. For example, many soft fruit varieties thrive in semi-shade, as do cut-flower bulb crops that finish most of their growing before the tree leaves come out.

Finally, it is worth noting that trees vary significantly in terms of growth habit and density of canopy. For instance, at Eastbrook Farm, the gooseberry plants under the apricot trees have failed miserably because, when the gooseberries want the most light, the trees are spreading with big leaves and little light finds its way through to the ground level. Conversely, the raspberries are planted under almond trees, which have a finer leaf and a more upright growth habit; these are thriving. When choosing trees, it is important to be aware of their height and spread to allow for shade planning.

Competition for Water

Shade and protection from wind reduce evaporation in hot weather. Tree roots and improved soil aid drainage in wet weather. However, there is a balance. It is possible for vigorous trees to compete for water with vegetable crops.

This had an influence on the design at Abbey Home Farm. We had looked at the results of agroforestry trials at Wakelyns agroforestry and noted they had experienced a negative impact on potato yields when potatoes were cropped in alleys consisting of willow (*Salix* spp) coppice during a drought year.[17] The yield impact was much greater than on winter wheat, for example. This makes sense, as the autumn-sown crops will be doing much of their growing before trees come into leaf and start shading and taking up water. It is worth noting, though, that even with the reduced potato yield the land equivalent ratio (LER) was still positive. In 2013, the potato yield at Wakelyns was reduced by approximately one third, but the overall LER was 1.36, meaning the system overall was 36 per cent more productive than if either the potatoes or the willow was grown on their own.[18] As vegetable growers, we needed to protect our potato yields; therefore, we reduced the amount of willow, ensuring it was only around the edge of cropping areas and not right in the middle. We are not saying don't grow thirsty trees like willow, as they have many excellent agroforestry qualities. Just ensure the distance between crops and trees is enough to offset this potential issue or know you have adequate irrigation capacity for your crops to counter the effect of thirsty trees in dry weather.

The competition between trees and crops will vary between sites, crops and years, so it is impossible to give specific guidance on how they will interact. For instance, although it sounds counterintuitive, in a very hot, dry year some leafy crops are able to thrive and grow only next to the trees because they benefit from the shade and reduced temperatures that the canopy provides.

Windbreaks

Most crops benefit from shelter from the wind. The increase in temperature and protection from the physical effects of gales are major factors for many growers when implementing agroforestry on their plots. However, understanding how windbreaks work can help to ensure that you maximise these benefits. We are trying to slow the wind, not stop it. A solid barrier makes the wind shoot up and over, creating turbulence and potentially more damage on the other side. A

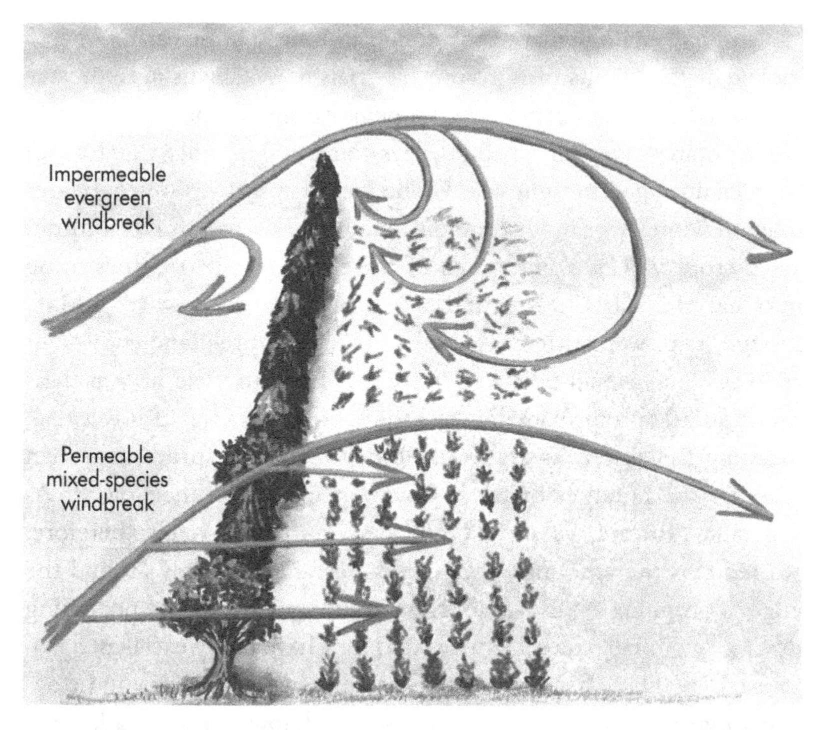

Figure 3.5. Windbreak effects.

well-designed windbreak will reduce wind speeds for about 10 times its height and potentially even up to 30 times.[19] Single rows of trees will provide some benefit, particularly if of mixed growth habit and species. However, by planting multiple rows with taller trees in the middle and shorter shrubs / coppiced trees on the outside we can direct the colder faster winds to higher levels, protecting the crops on the other side.[20]

Shelterbelts and tree rows also need to be maintained. How often have you seen a neglected hedge or coppiced row where the trees have grown upwards, leaving a bare stem and open area that tunnels the wind under the trees, increasing the risk to crops at ground level? Regular coppicing of some species within the row will ensure a mix of heights and habit within the windbreak. Similarly, including some evergreen species within your tree row can be beneficial, but even the bare frames of deciduous species will provide some protection over winter.

Beneficial Insects

Although we will look at the key factors that influence insect behaviour in greater depth in Chapter 5, here are some key basic principles.

Insect habitats need to provide food and shelter for all stages of insect life. The focus for supporting insects often concentrates on providing food during spring and summer. While this is important, there are a range of other requirements that vary through the seasons. Pollen is the key source of food for both pollinators and predatory insects during the growing season; and the other vital source of food for these predators is prey. Selecting a diverse mix of trees that provide a long season of pollen will help maintain larger populations of a range of insects, which in turn supports the predatory insects that we are so keen to encourage. Though it may seem odd, and is against the advice you find in more reductionist approaches, we also aim to grow tree species that are attractive to aphids and other pest species, such as caterpillars. With the holistic approach, we are not trying to remove the pests entirely from our system, otherwise what would the predators eat? Instead, we tolerate low pest levels in the knowledge that our range of predators will keep their numbers under control. Planting host trees near our cropping areas should encourage predators and help boost their populations. At Abbey Home Farm, wild cherry (*Prunus avium*), elder (*Sambucus nigra*) and spindle (*Euonymous europaeus*) trees have been included in the system design for exactly this reason.

Almost every part of a tree, from the tip of the canopy to the roots underground, can provide shelter for different stages of the life cycles of beneficial insects. Research carried out on predatory insects in Spanish orchards demonstrated that woody habitats are far more effective at boosting lacewing populations than flower habitats.[21] To optimise our trees for beneficial insects, we also need to understand the effective range of predators and pollinators from their habitats. Long-term farm-based trials in England have demonstrated that the effect of predatory insects significantly diminishes beyond a range of 50m (55 yds).[22] While this might make it tricky for larger arable farmers to incorporate enough biodiversity into their broadacre systems, for most vegetable growers, trees at 50m (55 yds) spacing still leaves plenty of space for cropping. For instance, tree rows 100m (109 yds) apart would allow insects to move 50m from each side and meet in the middle.

Crop Harvest Timings and Requirements

A very practical issue that can be overlooked is coordinating the potentially competing harvest requirements of both tree and vegetable crops. Let's look at timing first.

When is the crop going to be gathered? Will the trees cause any problems for vegetables grown near or under trees? A good example of this is growing lettuce crops for autumn harvest, if they are situated under or near to trees. The falling leaves from deciduous trees can drop onto the lettuces, slowing harvest and affecting the quality of the product. While a challenge for whole-head lettuce, this would be even more problematic for a mixed salad leaf production relying on a mechanical harvest. One option is to leave a large enough buffer between tree and crop, though this does not eliminate the problem totally. An alternative strategy would be to grow salad crops in the middle of the cropping area far enough away from the trees, or site them on the side from which the prevailing wind comes to reduce the risk of them blowing onto sensitive crops.

Will your trees get in the way of harvest machinery? Larger equipment, such as potato harvesters, might be tricky to manoeuvre around your agroforestry system. On a coppicing or pollarding design, we could time the cutting of tree rows for the winter before potatoes have grown to allow easy passage of the potato harvester, but in a system with standard fruit trees this would be tricky. With taller timber trees, there would be some difficulty when the trees were young, but once they had grown beyond the height of the equipment, with a clear trunk pruned for straight wood, this might not be an issue.

Now let's look at the tree crops. If we are growing apples, even at a relatively small commercial scale, we will want to bring some machinery alongside the trees, for instance, a quad bike and trailer. With a vulnerable vegetable crop next to the trees at the point of harvest, we'll end up either damaging the crop or taking twice as long to harvest the apples. For this reason, it may be necessary to ensure the vegetable bed next to apple trees is empty for apple harvest season. Again, this can be achieved through system design and crop planning, such as growing early harvest crops next to apple tree rows. This would be even more relevant if harvesting cider apples or perry pears from the ground mechanically. In fact, it is hard to see how you can incorporate cropping right up to these crops, so a grass buffer strip the width of the harvesting machinery is likely to be the only option.

Tree Habit

Tree habit is a botanical term that refers to the basic physical form and growing characteristics of individual trees. Trees can be bushy, conical, contorted, spreading, upright or weeping. In fact, all of these habits are present within the plethora of apple varieties available. Each habit can be either helpful or detrimental to an agroforestry scheme. A weeping tree will hamper vehicle access and understorey management but could provide good habitat and wind management. Whereas a conical or upright tree would lend itself to efficient access and understorey management but is likely to have a bare base that allows wind to pass underneath.

The speed at which the trees grow also affects system design. Some trees, such as alder (*Alnus* spp) or birch (*Betula* spp), grow very quickly in their first year, and might be grown as a nurse tree to help our longer-term species establish. Others lend themselves to forming a bush when coppiced. We might pick a tree species whose coppice cycle can be synchronised with the vegetable rotation. Willows (*Salix* spp) require coppicing more often than hazels (*Corylus* spp), and sweet chestnuts (*Castanea sativa*) require even longer. This timing is called phasing and is discussed in greater depth in 'Phasing', page 122. Mixing a range of species with different habits is especially important in windbreak design.

Rooting Behaviour

Just as trees can take on many forms above the ground, the same applies to how they behave underground. Rooting depth and rooting behaviour varies between species, which can have significant implications on interactions between trees and vegetable crops. If these interactions are well thought out, symbiosis can be realised, with nutrient recycling and positive water management benefiting vegetable crops. If not well thought out, trees have the potential to outcompete vegetables for nutrients and water.

Let's start by looking at the different types of tree root.[23]

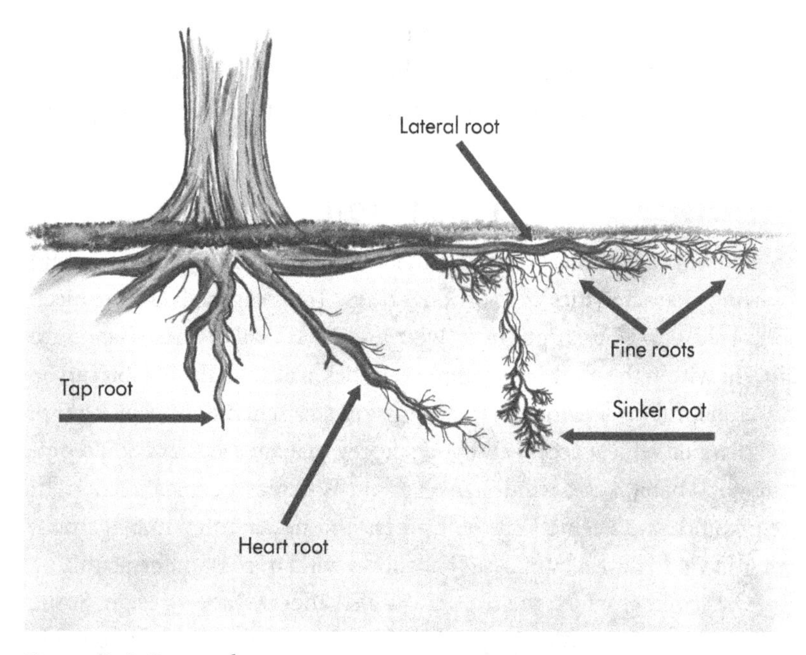

Figure 3.6. Types of tree root.

Typically, a tree seed will send out a tap root to get itself rooted quickly into the soil with enough depth to not to get blown over or washed away. However, in most cases, these tap roots don't keep growing down. Once established, the tree sends out a series of other types of roots to anchor itself and find water and nutrients. The fine roots grow from all the other types of roots and are where the tree will absorb water and nutrients. Crucially, they are also where the tree forms mycorrhizal associations. The horizontal extent and depth of the other types of roots will depend on species, soil and how we cultivate the land around them.

Most tree species have a root system that falls into one of the following three shapes.

Tap Root

The tap root is very rare in mature specimens but, as you can imagine, is very stable. From a cropping agroforestry point of view, this root system would work very well as the deeper central roots would not impact so heavily on neighbouring plants. Oak (*Quercus* spp), pines

(*Pinus* spp), almonds (*Prunus amygdalus*) and walnuts (*Juglans* spp) are some of the species that can form this root shape.

Lateral Roots

Lateral roots are the most common root system, with 80 per cent of tree species falling into the category, including birch (*Betula* spp), maple (*Acer* spp), pear (*Pyrus* spp) and willow (*Salix* spp).

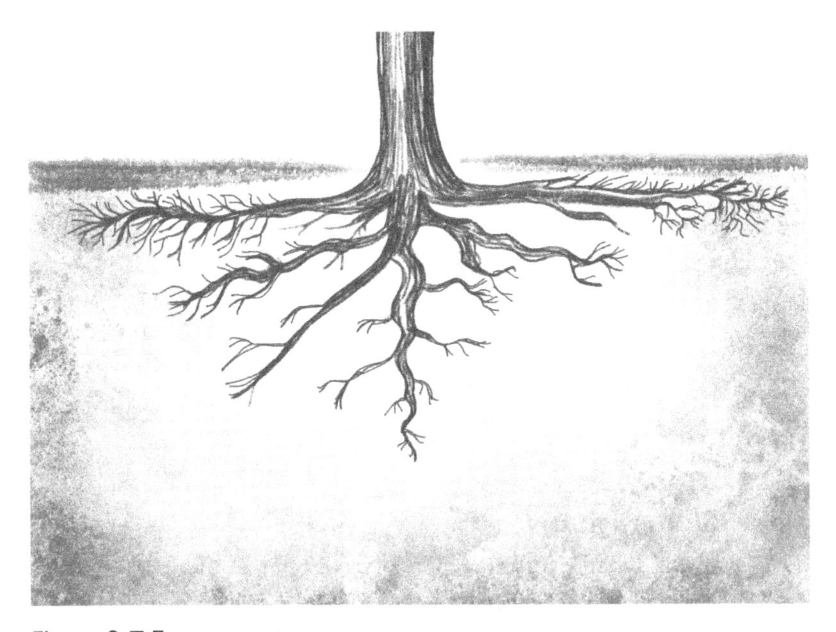

Figure 3.7. Tap root system.

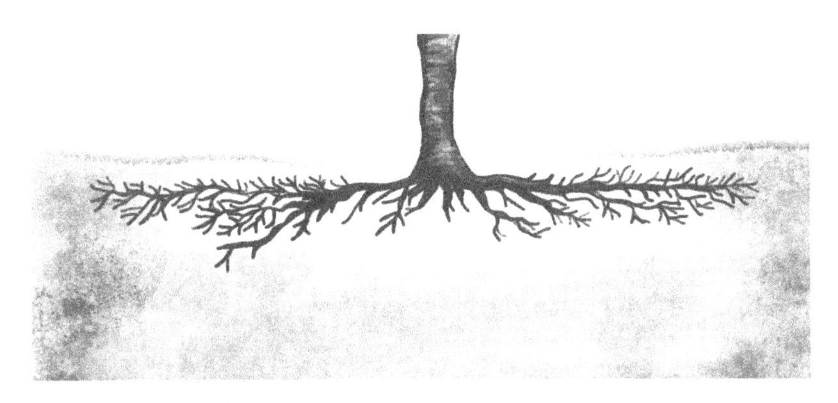

Figure 3.8. Lateral root system.

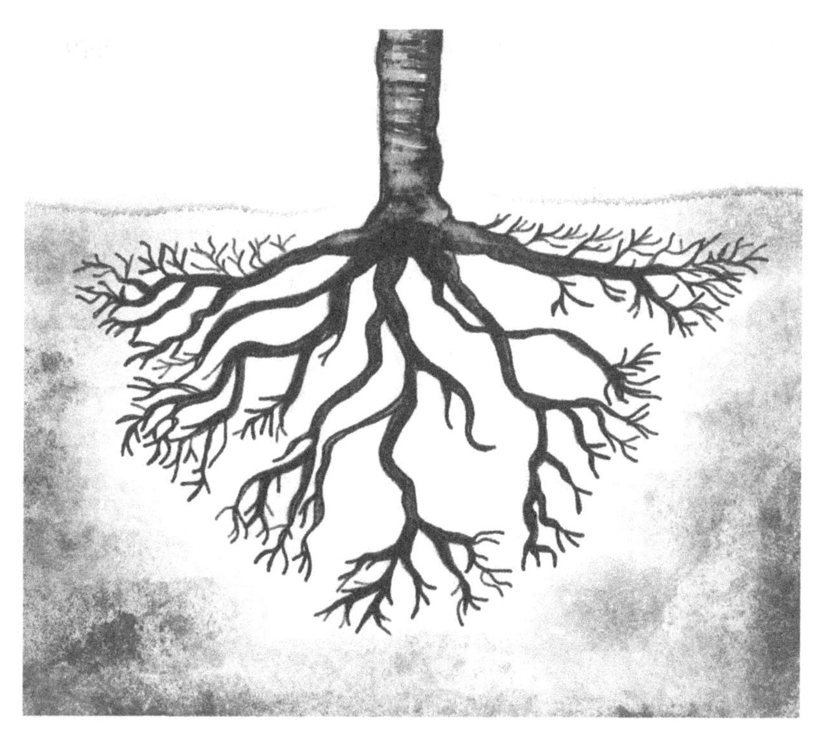

Figure 3.9. Heart-shaped root system.

Heart-Shaped Roots

This shape is what is often depicted in simplistic pictures of trees and their roots. It is actually not very common, but some oaks (*Quercus* spp), sycamore (*Acer pseudoplatanus*) and oleaster (*Elaeagnus* spp) have this form.

What we can see from these pictures is that, even in those trees with deeper root systems, there is still significant lateral rooting. And, importantly, many of the fine roots are concentrated in the surface roots because they need oxygen to breathe and live and that becomes scarcer in the lower soil levels.

There are some trees, notably many *Prunus* species, such as peaches and plums, that will send up suckers from those lateral roots. These will not only take energy from the main fruiting tree but will compete with nearby vegetable crops. These suckers need to be removed by cutting and root pruning. Some species, like blackthorn (*Prunus spinosa*) and black locust (*Robinia pseudoacacia*) are so vigorous in their suckering that we would not recommend them in vegetable-cropping agroforestry systems.

For farms with wetter soils looking to improve water infiltration, it is clear that some trees will do a better job than others. The ash (*Fraxinus excelsior*), for instance, though now problematic due to ash dieback disease, has a large root system with excellent ability to improve flow of water through the soil. Mixed-species stands also seem to outperform many monocultures for this outcome.[24]

Wrong Tree, Wrong Place

Although trees are broadly tolerant and resilient plants, they have their limitations. At Abbey Home Farm, we pride ourselves in growing a huge diversity of fruit and vegetable crops. We would dearly like to grow blueberries and to try growing citrus trees in our glasshouse; however, both of these require acidic soils to thrive commercially. We have mildly alkaline soils at the farm, so we have accepted that these two crops won't work for us. The same would apply to growing high-quality timber crops on an exposed cliff top. It is important to be realistic.

Although this book is primarily vegetable focused, many growers also keep some livestock, and there are a number of trees that can be toxic to animals. Often the risks are exaggerated and seen mainly in livestock that do not have access to a wide range of forage species. As an example, yew trees (*Taxus* spp) are widely avoided with farmed animals, but some anecdotal evidence shows that, in very small amounts as part of a varied diet, cows suffer no harm from eating some yew. If you have young children in your garden, then similarly there are some species that you might want to avoid.

There is a wealth of benefits to the soil with the 'no-dig' market-gardening approach and the regenerative farming approaches of no-till and min-till. Within these systems, there is no ground preparation during which you will attend to root pruning and management in a systematic way. These growers need to pay attention to their tree roots and crops, and build a way of coping into the system. If not, tree roots will have free run to invade the cropping areas at all depths, outcompeting the cash crop for water and nutrients.

So not only is tree choice on its own important but, as discussed throughout this chapter, if all of the interactions between tree and crop are not taken into account, trees can quickly have a negative impact on the crops around them.

CHAPTER FOUR

Identifying Objectives and Limiting Factors

Agroforestry is a toolkit that can solve a wide range of potential and existing problems, and can create new opportunities in vegetable production. We are convinced that these challenges will only increase with our changing climate, but similarly a shift in weather patterns may also open new cropping doors.

A thorough assessment of the site, as discussed in Chapter 2, should make you aware of site-specific challenges, such as a steep slope leading to soil erosion from rain, an exposed windy site leading to soil erosion from wind, or a site at the bottom of a valley prone to flooding. Maybe you have identified a complete lack of natural habitat on site for pollinators and predators. You might even feel at the time of designing your system that wind, floods and drought are not huge problems on your site, but we would urge you to allow for the intensifying nature of these over the coming years as climate change progresses.

Assets

Before designing, we also recommend that you list all your assets, such as current trees and available vehicles. Some will be obvious, but others that you might not consider (for instance, the availability of skilled labour) may have a significant impact on how you design and implement tree rows. Figure 4.1 is an example of a template we have used to help ensure we have considered everything.

Natural – especially existing trees and hedges, also rivers, ponds
Key species already present on site

Infrastructure – barns, irrigation, roads, access, utilities such as electric, gas and communications

Social – staff, community, customers

Farm element – arable, livestock, horticulture

Financial – capital available, access to grants/loans

Figure 4.1. The present picture of existing assets.

Objective Setting

What am I trying to achieve with my agroforestry system? At the beginning of Chapter 2, we talked about how every scenario is unique. This cannot be emphasised enough. However tempting and easy it may seem to copy an existing system, we urge readers to take *inspiration* from other designs but not to copy them wholesale. This is true not just because site and soil vary, but because of an array of other differences, such as grower skills and preferences, local community and business opportunities, or even availability of the right people or infrastructure.

Summary of the Design Process

Step 1: Understand your farm as it is now.

- Existing assets
- Inputs/outputs
- Barriers

Step 2: Set and prioritise your agroforestry objectives.

Step 3: Define broad structures – for example, rows and blocks – and how they will fit in your fields.

Step 4: Define final spacing and arrangement of trees and crops.

Step 5: Select the specific species and how many.

- Definition of functional groups – for example, nitrogen-fixing, coppice, large evergreen tree
- Species matched to functional groups

Step 6: Define establishment and protection requirements.

- Planting methods
- Weed management techniques
- Pest management techniques
- Livestock integration techniques (optional)

Step 7: Write up your design and plan.

- Drawings and maps
- Bill of quantities
- Establishment processes and calendar
- Management processes and calendar

As an example, we were talking recently with a grower keen to grow more short-rotation coppice and use the woodchip on his holding. As he researched costs and potential for mechanisation to move away from expensive chainsaw harvesting, he realised that the nearest contractor for harvesting was 200 miles away. The cost of buying the machine was also prohibitive, and so he adjusted his plan. For those with the capital to invest in that kit, there are clearly some contracting opportunities that could make the finances work, but that wasn't what he wanted to do.

While not everyone needs to go through a detailed objective-setting exercise, we do recommend looking at all the steps in Summary of the Design Process, page 49, particularly if you are relatively new to agroforestry.

Inputs and Outputs

We find this is a particularly useful stage of the objective-setting process. It forces us to look at the flow of materials through the farm through the tree lens. Some materials will present obvious substitution opportunities; for instance 'I am buying fence posts; I could be growing my own.' Others might not be so clear. Fertilisers, soil amendments, bedding, animal feed and energy are just a few of the things to which trees might be able to contribute. Try to list everything, even if it doesn't seem relevant. You can, of course, also include outputs that could come from trees, such as new products or even income from visitors. One farmer we worked with was planting trees to make his glamping business more attractive to guests. His plan included a wide range of fruit trees around each of the accommodation buildings that his patrons could pick during their stay. We've done an illustrative example in table 4.1 to get your creative juices flowing.

Having done a full site assessment and identified limitations and opportunities offered by your site, it's time to sit down and start the process of designing your system. When we designed our own agroforestry systems, there was a lack of information to inform our design processes. However, that is no longer the case. The sheer volume of information and options from all over the world can provide the opposite problem and quickly become overwhelming and make it very hard to work out where to start. Every tree will provide multiple outcomes. How do you find your way through? This is why the most important singular question you can ask yourself is, 'What exactly am I trying to

Table 4.1. Example Inputs and Outputs from the Farm

Farm Inputs	Farm Outputs
Compost Fence posts Diesel for vehicles and machinery Seeds Plastic for mulching perennial vegetables Timber for construction Bamboo for trellising crops Packaging Crop protection meshes Polytunnel plastic	Fruit and vegetables in box scheme Waste packaging – broken plastic crates, compost sacks Expertise – grower talking for farm walks, gardening clubs, school children Therapeutic space for community Venue for leisure or celebration Habitat for wildlife Employment and training

achieve?' Where you have multiple objectives, which most of us do, then try to prioritise them, either in terms of importance or urgency. Following, we have laid out a few of the most common objectives, but there will no doubt be others.

What Benefits Could Trees Bring to My Vegetable Cropping?

Reducing wind speeds, stopping water logging, managing soil erosion, reducing pollution of waterways, increasing pollinator populations, increasing predator populations, providing shade, recycling nutrients, fixing nitrogen and enhancing soil life and health. With so many benefits available to vegetable crops, it is important to understand that, although many trees will deliver a lot of these benefits in some way, if there are any particular benefits you are looking to achieve it is good to set this as a key objective at this stage. Setting a specific key objective rather than adopting a general 'trees are good for crops' attitude will help you tailor the design of your layout and your tree choice to maximise the effect of the chosen key benefits.

A good example of this is using trees for fixing soil nitrogen, a benefit that often comes up in discussions on agroforestry. While all trees should provide some nitrogen to the growing system, to focus on nitrogen inputs you need to specifically design the layout and tree choices around this objective. Trees would be nitrogen-fixing species, such as alder (*Alnus* spp) or sea buckthorn (*Hippophae rhamnoides*), so the layout would also have to be designed accordingly. You might also design a system that cuts the trees when green to add into your other growing areas. You would also need to understand the impact to the soil biology of adding this material.

If reducing wind speeds is the objective, then the spacing of trees within the cropping area might be less dense, but building in a mix of species might be more important. There will be general benefits to pollinators and predators by introducing trees into cropping areas, but if this is to be a key objective of the system, then choose trees that provide the right habitat for the species you are trying to support. For instance, the effect of a long season of pollen for flying insects and undisturbed perennial habitat to support all stages of life can be magnified by applying the same approach to tree understorey. An understorey of rhubarb planted through plastic mulch will produce more income but support relatively little biodiversity, while a perennial wildflower understorey would be hugely beneficial for pollinators and predators. The layout design plays a key role in achieving this objective, linking habitats, providing wildlife corridors and ensuring regular effective habitat across the whole cropping area.

If the desired objective is to reduce soil erosion on a site with a steep slope, then the design of the layout would want to position trees across the slope, to prevent soil from being washed down the slope. However, if this put the tree rows in an east-to-west orientation, the design would need to cater for areas of constant shade on the north side of the tree row. This demonstrates that objectives can have negative effects if not accounted for in the design process.

Do I Want to Introduce New Crop Lines?

For vegetable growers, the idea of introducing new edible crop lines is not a daunting prospect. In fact, it is often one of the most attractive aspects of agroforestry to growers, many of whom have existing outlets and will frequently be buying in fruit to supplement their vegetable offering. Of course, as with any new vegetable crop, we need to understand our planned crop type and make sure we know what is required to successfully grow, protect, harvest, store and process the crop and finally sell the end product.

Labour, materials, infrastructure, equipment, time and cost are all variables that must be considered at every stage from planting through to mature crop. In the case of apples, they can take 10 years to reach maximum yields, a much longer time scale to plan compared to vegetables.

In addition to edible tree products, there is a whole range of non-edible opportunities that, as a vegetable grower, you might not have considered. Timber, woodchip, medicine, floristry, fence posts and bean poles are amongst the many non-edible crops that can be produced from trees. We will look at these more in Chapter 6. Broadly, the questions to ask are the same as for edible crops.

Do I Want to Reduce Farm Inputs and Produce Them on Site?

In recent years, the price of fossil fuel–reliant farm inputs, such as artificial fertilisers and pesticides, has increased to unsustainable levels for many farmers. Agroforestry, if designed correctly, offers good opportunities to produce or reduce these inputs on site, lowering costs significantly and building resilience into your system by being in control of your inputs and not at the mercy of world events and markets. For vegetable growers specifically, there are a few potential outputs from trees that might help to reduce external inputs.

Trees can be used to make fertilisers. Though a holistically functioning garden shouldn't need many additions, there are times when certain crops need a boost or a certain mineral. Often those nutrients will be present in trees, and we can apply a liquid fertiliser or soil amendment made from home-grown material. We recommend Nigel Palmer's *Regenerative Grower's Guide to Garden Amendments* for those interested in this topic.

There is a myriad of uses for composted woodchip in horticultural systems, including mulching on paths, around trees and in no-dig systems. You can also produce compost not just for applying to cropping areas but even for making your own propagation substrates for seedlings and transplants. Propagation substrates are an important area of focus if we are to stop our reliance on peat-based composts that contribute to carbon release and habitat destruction.

We can also grow, harvest and apply fertiliser directly to growing areas without the need for composting in the form of ramial woodchip (material from branches less than 7cm [3 in] in diameter). This is particularly useful if we have suitable coppiceable trees within our production systems, as we can reduce the time and cost of moving material about and tying up space for composting.

Although we are organic growers and do not use any pesticides, for those still relying on chemical interventions, trees and their understorey are a very powerful tool for managing predatory insects to levels where artificial pesticides can be reduced or even eliminated.

Can I Use Trees to Adapt to Climate Change?

At Abbey Home Farm we are often asked if we have seen an increase in yields since we planted trees; our response to this question is normally, 'Not yet, trees are a long game.' However, our view of the long term has changed since we planted the trees. We expected there to be a long-term positive effect on yields, but we now see that positive effect in a different way. Over recent years, the effects of climate change have accelerated at farm level. We find ourselves as farmers on the front line. Extreme weather events are becoming increasingly frequent and varied. Now we see the trees as a key tool to helping to maintain yields rather than to increase them, and a vital tool for adapting to and mitigating the effects of climate change.

Importance of Diversity

With so many different outcomes possible, you can achieve a number of key objectives at the same time. To allow this to happen, build diversity into your design. At Abbey Home Farm, we grow 90 different fruit and vegetable types, and it is literally impossible to get a sequence of weather that would allow all 90 to crop to their maximum potential. However, at the same time, the diversity means that, whatever the weather is doing, a significant proportion of crops will be cropping strongly.

The same approach can be applied to tree crops, whether it be a variety of top fruits or a mix of edible and non-edible crops. A coppiced tree is far more weather resilient than a speciality fruit tree. For example, the coppiced tree may not earn as much but will fruit more reliably. As with vegetable crops, pest and disease are ever-present challenges with tree crops and perhaps the biggest challenge with top-fruit production. Diversity is one of the most effective ways of disrupting species-specific pests.

Darwin recognised the power of diversity and there is more recent research to back up his theory that species diversity increases the overall productivity of a system.[1] As humans, if we are building a team to deliver a house-building project, we want a mix of skills. It's no use having nine plumbers and no electrician or carpenter. So it is with plants, their rooting systems, nutrient demands, and how they fill ecological niches and interact with other animal and plant species. Each plant will be slightly different, which means they support rather than compete with each other.

There is a system of expressing this increase in productivity in agroforestry systems called *land equivalent ratio* (LER). It works on the basis of how much extra can be produced by the two elements (tree and crop) when grown together on the same piece of land compared to how much you would get from growing a monoculture of either of them. For instance, let's say you have one hectare from which you could expect either 50 tonnes of potatoes or 50 tonnes of apples. By integrating the two crops and planting the one hectare with a half hectare's worth of each, you might get 30 tonnes of potatoes and 30 tonnes of apples. The result is 60 tonnes of output from the one hectare rather than 50 – 20 per cent extra production. Your LER would be 1.2 (or 60 ÷ 50).

Limitations

Although we are always excited about the potential to be realised by agroforestry, let's inject a dose of realism into the design process to make sure we can actually deliver on that potential. There are always commercially limiting factors to consider. These limitations may not alter the actual design of the system, but they might influence the planting and establishment time scales.

The most likely barrier to being able to quickly implement any new planting is money. Planting trees costs. As well as the actual tree, there's stakes, ties, guards and fencing to buy. There is then a cost for maintaining the tree until it becomes established: pruning, weeding and restaking, perhaps, to mention a few. In the UK, there is money available for to help with planting trees, supported both by the government and from private funding. However, it is important to realise that funding streams can be restrictive and unpredictable.

Finding suitable skilled labour for planting and tree maintenance is also a challenge. This might be less crucial in vegetable systems where growers already have a good understanding of plants and growing, but nonetheless, understanding how to prune a range of different trees with multiple outputs or learning how to train a tree to grow in a particular form can be daunting. That labour also needs to match up seasonally with other jobs on your land. This can work out well – a good proportion of work on trees happens over winter when vegetable-cropping tasks are much reduced, but the tasks still needs to be planned. There are some real opportunities for new entrants or collaborations between sectors. Aspiring growers with no land could manage the trees on a number of holdings or come to a share farming agreement with you so that you can continue to focus on the vegetables while still benefiting from the output of the trees.

Sometimes, even when you have both the money and the right people, it can still be hard to fit everything in. The classic example of this is winter tree planting. In the UK, we used to be able to reliably plant trees from November to March, safe in the knowledge that the trees would remain dormant. With climate change, we are finding that nurseries often will not lift bare-root trees before December, and planting them much after mid-February is risky. Using cell- or pot-grown trees can expand this planting window, but it still concentrates what can be big jobs into short time slots.

Knowledge and experience are the final major barriers, in our experience. We have decades of growing knowledge between us, but until recently this had been limited on the tree front to a very small number of common species, and mostly apples. Even with more understanding of trees, the pair of us still need to work out what works well in our specific systems and on our specific land. At Eastbrook Farm, we have deliberately slowed down our programme of tree planting, not because our long-term ambition has changed but because it takes so long to know whether we are succeeding. If we grow a vegetable crop and it fails, we know within months or at most a year. We can then decide not to grow it again or to change our method of production. With trees, however, we may not really know if they are productive and happy for 10 to 15 years. We are only eight years into our planting, so to rush in and replicate our first efforts – or conversely to ditch something that struggled a bit to start with but might now be thriving – feels premature. Planting trees is a long-term game, so we aim to take it slowly and build on our experience at a tree's pace.

CASE STUDY
Martyn Bragg, Shillingford Farm

Martyn Bragg runs a 420-acre fruit and vegetable farm in Devon. He sells his produce locally through his own award-winning box scheme and at farmers markets. Some areas of the farm were left exposed after trees were removed in the 1960s. This made the land vulnerable to strong winds and soil erosion, with the steep slopes especially susceptible. Martyn knew he could sell fruit locally, so planted his first alley cropping system in 2002. In addition to increasing his product range, Martyn was keen to support biodiversity and reduce the wind affecting his vegetable crops. The trees also provide interest for customers who come on farm walks and pay for farm tours.

SYSTEM

Martyn's first design was at 12m (39 ft) spacing, planted by rolling out a 2m-wide (7 ft) strip of woven plastic mulch to control the couch grass. The trees are planted on a south-facing slope with the rows oriented north to south. The bottom of the field is wetter and frost collects in that section, so trees were not planted there. Once the system matured, Martyn realised that the alleys were not wide enough, making it tricky to use machinery and manage the understorey. He was also concerned about the amount of shading on his vegetables nearest the trees. After 10 years, there is now only space for three beds (down from four originally) between the rows, despite regular pruning of the trees.

Martyn learned from this and his final planting in 2013 of apple (*Malus* spp), pear (*Pyrus* spp), common mulberry (*Morus alba*), common hazel (*Corylus avellana*) and common alder (*Alnus glutinosa*) trees was set up with trees 28m (31 yds) apart in 4m (13 ft) rows, giving a 24m (26 yds) alley for vegetable production. Much of this second planting was on land already sheltered by a hedgerow that Martyn had let grow high along the northern edge of the field to act as a shelterbelt. This protection enabled his trees to establish more quickly. He also planted a couple of rows of willow for biomass. Martyn also decided to go for a less vigorous apple rootstock in his second planting, using MM106 rootstock rather than MM111, to make picking easier.

Martyn controls the understorey in two ways: by strimming and mowing, both pre–apple harvest and in the spring. He also uses chickens under the trees, with the eggs feeding into his retail outlets. He has been using a local orchard expert to prune the trees, but they are now retiring so Martyn has trained one of his farm team to take this job on.

KEY OUTCOMES AND LESSONS

Martyn believes that his alley cropping system delivers greater profitability through increased vegetable yields and supplementary fruits, nuts and fuel. As well as the fruit and nut sales, he has installed a biomass boiler to heat his farmhouse and buildings. The increase in shelter helps to warm the soil earlier in the spring, so the vegetables can be harvested sooner, which is essential for Martyn's continuity of supply within the box scheme. Shelter also helps to limit the negative impact of drought, by reducing wind speeds and improving crop water efficiency.

Vigorous rootstocks produce more apples but are harder to harvest. Martyn's pears are grown on Quince A rootstock, as he feels pears are trickier than apples and need this added vigour. He has almost no apple diseases, due to increased air movement from wide row spacing. Grass and brambles under trees are a bit of a problem but great for biodiversity. Martyn suggests growing very early apples if you are selling directly, as being the first person to have new season apples is a great marketing point. He also recommends having some varieties that store well to extend the season at the other end. Martyn trialled using a biodegradable mulch to control weeds, but that degraded too quickly so he wouldn't recommend it.

CHAPTER FIVE

Trees and Nature

Natural pest management (NPM) is a philosophy that involves tolerating and managing pest species instead of eradicating them. This is not a 'let nature take care of it all' approach but rather a harnessing of the power of complex biological systems to optimise yields and remove the need for damaging chemical inputs. Growers able to observe and understand in depth the crop, potential pests and the environment will get the most from this approach. NPM strategies focus on harnessing inherent strengths within ecosystems and directing pest populations into acceptable bounds rather than eliminating them. This approach avoids undesirable short-term and long-term ripple effects and will ensure a sustainable future.[1] NPM principles apply to cropping systems over an extended spatial and temporal scale rather than to individual crops.[2]

In short, NPM is all about farmers harnessing the powerful tool that is biodiversity. Integral to healthy biodiversity is a diverse range of complex habitats. On most farms in the United Kingdom for instance, these habitats will include the soil, grassland, wetlands, rivers, hedgerows, scrubland, woodland and cropping areas. In all these habitats, trees play a unique and very important role, with the notable exception of cropping areas, which currently are mostly devoid of trees. After the Second World War, hedges were removed to make fields bigger and enable greater efficiency with larger machinery. Though small in comparison to fields in the US, Eastern Europe or China, some fields in the UK are now more than 300 acres.[3] This increase in field size is a phenomenon in horticultural as well as broadacre crops. As those trees and hedges are removed, biodiversity decreases and our ability as growers to naturally manage pest and disease problems also reduces.

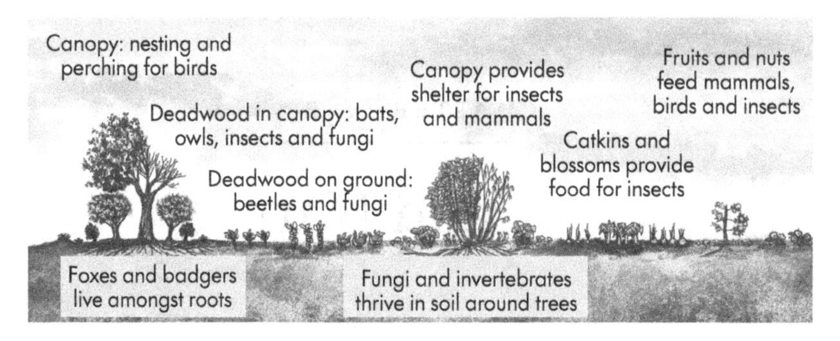

Figure 5.1. Trees as habitat.

While we recognise and cherish the very important role that complex mixed-tree habitats such as woodlands and hedgerows play in biodiversity, which will be discussed in the following, we should also not forget the important role of individual trees in the modern farming landscape. Lone trees and small isolated copses within vast monocultures can be the last islands of biodiversity and a crucial refuge from modern farming practices. This chapter will explore how welcoming trees back into the cropping landscape will provide a crucial role in protecting our crops.

Trees as Habitat

A habitat supplies everything an animal needs to find and gather food and water, select a mate and successfully reproduce. The perennial nature of trees gives them a crucial role throughout the seasons. Blossom, shade, fruit and shelter provided by trees are key parts of many animal, plant and fungal life cycles. All parts of a tree play their own roles as habitat: roots, trunk, bark, branches, leaves, flowers, fruit, nuts. Due to the huge variety of habitat and opportunities within a tree, many animals have based all or specific parts of their life cycles around trees. Let us look in more depth at how different life forms rely on trees, before looking at how to harness the life in trees to protect our crops.

Fungi

Over 80 per cent of fungi in the United Kingdom are associated with trees. The interconnectedness of tree and fungi life cycles is a rapidly expanding area of understanding and knowledge. From the top of their canopy to the tips of their roots and beyond, fungi play an integral role in the life of all trees and vice versa. While the fantastic visual array of fruiting bodies – mushrooms or toadstools – is often what springs to mind when fungi are mentioned, these are but momentary glimpses of these fascinating organisms, the true magic of which often hides out of view in the form of a mycelium, a lace-like network that makes up most of the organism. This book is not the place to celebrate in detail the almost infinite variety of fungi and the many different complex roles they play in a tree ecosystem. All we need to understand are the key functions that fungi provide in relation to trees and how that, in turn, can benefit our horticultural crops.

Decomposers

Fungi play a key role in recycling organic matter from dead plants and animals. For woody material, particularly, they are the primary organism able to cope with breaking down the tough lignin structures found within wood and bark. Fungi also need less nitrogen to be able to carry out that decomposition, compared with bacteria, the other major player in organic matter recycling.[4] Once the fungi have started the process, other organisms can take the partially digested material and further deconstruct it into forms suitable for plants to take up through their roots.

Nutrients

Many fungi form symbiotic relationships with trees and their roots, sourcing, extracting and transporting nutrients that they then exchange for sugars with the tree roots, a process vital to the survival of tree, fungi, and other plants growing in the area. Fungal mycelia can stretch for large distances. The largest organism in the world is thought to be an *Armillaria* fungus living in the Blue Mountains of Oregon, covering 10 square kilometres (4 mi^2). Though most are nowhere near that size, as growers we can harness that power by encouraging fungal populations to deliver nutrients from some distance to our cropping plants.

Defence

Some fungi take this symbiotic relationship to the next level. The fungi receive benefits from the tree roots in exchange for defending the roots against attack by other fungi, nematodes and invertebrates. There is extensive film footage of a fungal hypha forming a noose, which it uses to snare and devour a passing nematode.[5] There is also evidence that messages might travel through fungal and root networks between plants, giving them advance warning of pest attack, for example. [6]

Invertebrates

There are more than 2,000 different invertebrate species in Britain alone that are dependent on decaying wood to complete their life cycles. An English oak (*Quercus robur*) can support around 285 different insect species, and a white willow (*Salix alba*) more than 200 different species. The tree habitat, both living and dead, provides a number of services to these invertebrate populations.

Trees furnish food to all life stages for many invertebrates, whether that be dead wood, sap, foliage, live wood, pollen, nectar, lichen, fungi or other animals. In the farming landscape, trees are often one of the first sources of pollen in the year, a crucial food source for many adult insects. They provide shelter from weather and refuge from predators, whether to egg, pupae, larvae or adult. Many invertebrate life cycles are, in fact, inextricably linked with other plants, fungi, invertebrates and other animals in complex symbiotic and parasitic relationships. In turn, invertebrates are often crucial to the life cycle of the trees in which they live. Above ground, they pollinate many species, and for most species there are invertebrate predators of tree pests. Underground invertebrates also play a role in the rhizosphere (the complex habitat around the roots), sourcing, processing and exchanging nutrients with the tree and providing defence against attack. Many invertebrate species are key recyclers, breaking down dead wood and leaf litter into organic matter, which can then be digested by smaller organisms.

Birds

For a group of animals that specialise in staying off the ground to keep safe, there is no more important habitat than the tree. Birds nest in their twigs, branches and trunks, safely away from most ground-dwelling predators. They feed on the nuts, berries and seeds produced by trees, as well as eating the myriad animals that live in and around trees.

As is so often the case in nature, this is not a one-way relationship, and birds help trees to reproduce by dispersing seed over wide distances. In fact, some seeds will only germinate after passing through the gut of a bird. While birds can be crop and tree pests – for instance, bullfinches eat fruit buds and lots of species eat commercial fruit crops – they can help defend the tree by keeping pest numbers at tolerable levels, such as caterpillars that might otherwise defoliate a tree. Certain species of trees will even host apex bird predators, such as goshawks and sparrow-hawks, that can reduce the numbers of pest species. As we will see time and again in our striving for true natural pest management, balance and holistic planning are the key.

Mammals

More than 30 different mammals in the United Kingdom rely on trees or areas of trees as vital habitat. Badgers, otters and foxes live around their roots, dormice amongst their stems, bats in their hollow trunks, and deer and wild boar shelter from the weather beneath them. As with the other groups of organisms described previously, tree habitat provides both shelter and food to a wide variety of mammals, and again, in return, the tree gains benefits such as seed dispersal and defence from potential threats. Many mammal species can quickly become a problem in farming situations, particularly when trying to establish trees. Deer, voles, rabbits and squirrels are all major pests of young trees and keeping their numbers in balance is not always easy, especially since in the UK there are no natural predators of deer and few of the grey squirrels. However, once the tree has grown large enough, most species can fend off a degree of browsing or damage.

Maximising the Benefits of Tree Habitats

We can see that the habitats associated with trees are highly complex and dynamic. While not able to comprehensively map all possibilities, in the following we describe some of the key interactions between trees, wildlife and crops that we have observed in our own silvohorticultural systems. Each site will be different, with topography, climate, soil and geography all playing a part, as it will in your individual farming system. The key is to observe and ask questions. Which insects are visiting trees and at what time of year? Are there times of year when there is no source of pollen, and does this coincide with a pest attack? If you have a major pest for one of your crops, what can you do to encourage predator populations at crucial points in the growing season? In the initial stages, you might need to assist nature by putting up bird and bat boxes or growing some annual pollinators alongside tree species. As your system develops complexity and resilience, the aim would be to reduce the need for these assists.

Pollen

In mid- to late February as we walk the fields and look for opportunities to start some ground preparation for early crops, such as onions and potatoes, wildlife is also just starting to show itself. As the days lengthen and temperatures slowly rise, the first of the insects venture out from their sheltered overwintering spots in woodlands and hedgerows, looking to feed and to lay their first eggs. Many of the predatory insects that feed on other insects, such as aphids, use pollen as an alternate source of protein to power their flight while looking for insect prey to eat or places to lay their eggs. Early in the spring, pollen is hard to come by in the countryside and one of the earliest sources is on the catkins of willow (*Salix* spp), common hazel (*Corylus avellana*) and common alder (*Alnus glutinosa*). These three species will produce pollen in succession from February to April, depending on geography and aspect, and for this reason they have formed a major part of our agroforestry planting at Abbey Home Farm. Amongst them we have a number of 'goat' or 'pussy' willows (*S. caprea*) and common hazel (*C. avellana*), which are the first to produce catkins. As soon as they develop pollen, during every sunny spell through the day they are covered in a variety of insects, both pollinators and predators. This knowledge is an important tool in maintaining and building insect numbers early in the season, even more so because populations of aphids generally increase faster in spring and predatory insects can lag behind, as they prefer slightly warmer temperatures. So, any techniques to assist early development of beneficial insect populations are vital.

By integrating these trees within cropping areas, we aim to keep these predatory insects close to our crops where they can easily foray out over our crops looking for aphids, caterpillars and other pests. It's especially important in spring when weather can change quickly, with trees providing substantial and convenient shelter from wind and rain, and keeping predators close to our crops where they can take advantage of short breaks in the weather to feed without risk from wind and rain.

A wide variety of insects play a crucial role in pollinating most human food crops. Nearly 80 per cent of plants in Europe are insect pollinated.[7] Many people think of honey bees when pollination is mentioned, and, while true and useful, one of the first visitors to the pussy willow (*S. caprea*) catkins each spring are actually hungry bumblebees. There is plenty of evidence that honey bees are not the most effective pollinators, as solitary bees and bumblebees often visit more flowers and transfer more pollen between plants.[8] Increasing pollinator numbers can help to increase yields, for instance of crops like broad beans,[9] tomatoes[10] and courgettes.[11] Even wind-pollinated trees like elm (*Ulmus* spp) and oak (*Quercus* spp) can provide vital food for insects, both in the pollen and in the honeydew excretions left behind by feeding aphids and other sap-sucking species.

It's not all about the bees, though. One of the most important groups of early pollinators, especially in a protected cropping environment, is hoverflies. There are more than 200 different species in Britain, and not only do they pollinate flowers but they are also one of the key aphid and caterpillar predators in horticultural systems. One larva can eat up to a thousand aphids before pupating after four weeks. The succession of pollen from trees in our silvohorticultural system will continue to provide this crucial alternate food source for insects from the catkins of February, through the apple and wild cherry (*Prunus avium*) blossoms of March and April, followed by the elder (*Sambucus nigra)* flowers of May and finishing with spindle (*Euonymus europaeus)* flowers in June. This provides a valuable food source for insects for five months and coincides with when our crops are at their youngest and most vulnerable to predation. After the tree pollen season runs out at the end of June, wildflowers and grass take over as a source of pollen. However, it is not long before the berries and fruit of cherries, apples and elders start to provide food from the trees again. This example demonstrates the need to think beyond either food crops or 'biodiversity' plantings of flowers and trees. To maximise our NPM impact, we need to fully understand and integrate the full range of crop and non-crop plants on the farm. When seen in this light, giving some of the farm to wildlife rapidly changes from being seen as a luxury to an essential tool of farm success.

Live Wood

The living wood of trees provides many services for biodiversity. Trees are a key source of shelter from weather for a variety of animals, whether that be in the branches, under the branches or, for smaller invertebrates, in the bark of trunks and branches. Allowing biodiversity to remain in the middle of cropping areas, and not forcing a retreat to the cover of nearby hedges and woodland, also allows for much more effective crop protection service.

Providing cover during bad weather spells in the growing season is one way the shelter of living trees is of service to the vegetable grower. A key predator of aphids on farms is the ladybird and its larvae. When winter approaches, ladybirds leave exposed cropping fields and head to more substantial habitats to escape the intensity of winter weather. Often it is in the bark of trees and old leaves that ladybirds can be found in the winter. Allowing predators, such as ladybirds, to overwinter on trees within cropping areas means they are much quicker to provide control against pests the following spring when the aphids return.

Another benefit of having living trees amongst one's cropping areas is welcoming birds back into the farming landscape, an absence iconically articulated by Rachel Carson in her famous book, *Silent Spring*. Not only are birds drawn to trees for the host of insects attracted in by their pollen as discussed above. They also make use of the opportunities the structure of the trees provides, as a place to both nest and raise young, and this, in itself, can have a huge positive effect on crops. One blue tit chick can eat up to a hundred caterpillars a day while still in the nest. An average nest will have 10 chicks, so one nest of tiny songbirds will require a thousand caterpillars a day to sustain it. With many nests of many different species, it becomes clear the powerful protection service birds can provide in controlling problem insects.

It is not only insects that birds will control from their vantage point high in the branches; birds of prey, such as kestrels and buzzards, are also able to survey the surrounding area. Kestrels are vole-hunting specialists regularly seen hovering over our crops, while buzzards are opportunistic and prefer easy carrion but will take small mammals if they get a chance. Although neither are rabbit specialists, they definitely influence rabbit behaviour in our cropping areas. The presence of birds of prey hovering, cruising over and perching to overlook crops means that rabbits don't dare venture out into the exposed open parts of the fields where they are a long way from cover. So while our local raptors do not necessarily catch a lot of rabbits, they limit them to the edges of fields and beginnings of crop rows. This behaviour demonstrates one of the key principles of NPM, which is managing a pest to acceptable levels of damage. We are not looking to eradicate caterpillars and rabbits, but we do want to keep their populations at tolerable levels.

Dead Wood

Dead wood plays an often unseen and uncelebrated role in fuelling ecosystems. It is the basic food source at the centre of many food webs and provides shelter and habitat for many animals even before it has fallen to the ground. Dead branches still attached to trees and whole dead trees provide an amazing variety of shelter for many useful predators. Owls and bats are both common residents of dead, old trees, and, while the former is a known hunter of rabbits and small mammals, bats are often missed as a potent controller of agricultural pests because they are mainly active at night. They are generalist eaters and have been recorded eating a variety of different agricultural pest species.[12]

Invertebrates that rely on dead wood are called saproxylic. More than 2,000 species of invertebrate in the UK alone are saproxylic.[13] The holes in dead branches and trunks offer refuge for smaller animals: social wasps build nests in them, while solitary wasps use them, too. Both kinds of wasps will feed on aphids and caterpillars. In fact, the group parasitoid wasps is one of the most effective predators of both caterpillars and aphids, often brought in as a biological control for protected cropping environments for exactly this purpose. Many hoverfly species, another key aphid predator, use the damp environment of standing dead wood to complete parts of their life cycle.

Dead wood both in the canopy and on the forest floor provides habitat for another important group of animals. One in four species known to science is a beetle, and they make up 40 per cent of all insect species, 25 per cent of all animal species.[14] In the UK alone, 650 species of beetle directly rely on dead wood to complete their life cycles. They feed on dead wood and the fungi associated with dead wood, and others are hunters that feed on the smaller grazers of dead wood. It's these active hunting species of beetle that are of most interest to the horticulturalist. Ground beetles and their larvae are vital predators in the farming landscape, feeding on slugs, slug eggs, caterpillars, aphids, cutworms and leatherjackets (crane fly larvae). They can also eat seeds and one study estimates that they can reduce weed burden by up to 30 per cent by eating their seeds.[15] The adults are mainly nocturnal and spend the day hiding away in the cover of permanent grassland, dead wood and leaf litter. This is also where they lay their eggs and where their larvae live. When harvesting carrots and potatoes on our farm, we are always amazed at the sheer number of beetles that thrive within an organic farming system. Trees themselves are an important part of beetle habitat; however, what is underneath the trees is equally important for ground beetles. Wherever possible dead wood should be left alone under trees, as this is crucial habitat for many beetles.

Roots

Although out of sight, tree roots play an important role in protecting your crops. Up to 25 per cent of the carbon absorbed by a tree through photosynthesis is released from the roots in the form of sugars. These root exudates are a crucial food source for a variety of soil life. One of the major consumers and mobilisers of these root exudates are fungi, and, in particular, symbiotic fungi called mycorrhizae. Many of these fungi form species-specific symbiotic relationships. However, there are also generalists that hook up with many different species of plants and trees.

All crops have the ability to defend themselves but this is greatly enhanced when they are growing in a vibrant healthy soil packed with symbiotic fungi. The best defence for a crop is simply to be a healthy, vigorous and competitive plant. The ability of fungi to mobilise nutrients for plants plays an integral role in this process. Fungi also play another role in protection as they can actively defend the plant's roots from attack.

Larger mammals, including predators such as foxes, polecats, stoats, otters and weasels, regularly make their homes amongst tree roots, and all of them can play an important role in maintaining rabbit and small mammal populations at acceptable levels.

Bringing Biodiversity onto Farms

Having explored the valuable role biodiversity can play in protecting our crops and identifying how trees support this, we now need to understand how wildlife populations move around the wider landscape. Then we can make use of this understanding to bring that biodiversity right in amongst our crops.

Permanent Habitats

According to the United Nations, 86 per cent of endangered species are at risk solely due to agricultural practices. Modern farming has decimated habitats and reduced what were mind-blowingly complex, diverse ecosystems to isolated pockets across our countryside. These refuges exist in the form of wildlife reserves, woodlands, copses, field boundaries, hedges, roadsides, meadows, coastlines, rivers and wetlands, and are important stores of biodiversity. They need to be sensitively managed and, wherever possible, allowed to expand with any human interference kept to an absolute minimum. It is in these permanent habitats that complex food webs develop from fungi and plants right up to apex predators such as raptors and foxes. These habitats are a wealth of biodiversity resources for sustainable farmers.

Animal populations are highly dynamic and do not remain within the confines of your field or farm. Many birds, insects and mammals will retreat to the shelter of dense woodland or scrubland for winter to find cover from predators when in autumn the leaves fall and plants die back. Identify these habitats in the landscape in relation to your farm and visualise how you can encourage biodiversity from these refuges back to your cropping areas during the growing season, or perhaps even persuade them to stay in and around your cropping areas all year round. In reality, to achieve this we must establish complex permanent habitats in and around our crops, and this takes time. So even if we aim to start establishing these permanent habitats, we still need to access the biodiversity in established permanent habitats in the interim.

Wildlife Corridors

Wildlife corridors can provide the critical link between your farm and permanent complex habitats that might be some distance away. Open farmland is generally a dangerous place for wildlife. They are exposed to weather and predators. Safer ground surrounds and runs alongside fields, less disturbed by humans and with thicker cover. The longer the distance wildlife can travel out of sight the better, especially if they can find food enroute. Identify and map wildlife corridors near your site. On the ground exploring works well, as does using aerial photos and

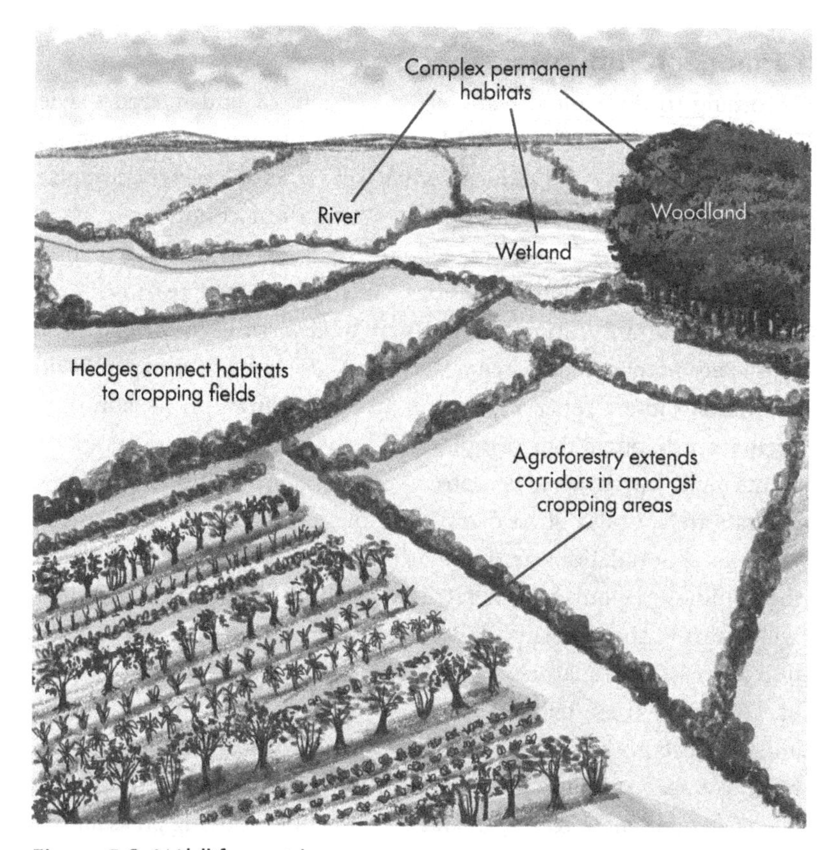

Figure 5.2. Wildlife corridors.

satellite images. Rivers, hedges, canals, steep valleys, railway lines (both in use and decommissioned), coastlines, roadside verges, drainage ditches – the list is endless and many of them involve trees.

Having identified which of these landscape features link your farm to the permanent habitats, try to engage with other local farmers and landowners to share knowledge of the value of these wildlife corridors and help to protect and enhance them. Within your own farm, you can start to manage wildlife corridors to bring animals right up to your cropping areas throughout the seasons. For our farm, the hedgerow is particularly important. Even though it is typical of the British landscape, it does not always get the cherishing and celebration it deserves. The hedgerow is the superhighway that allows animals of all sizes to move away from the threats of modern farming. Looking at Fig. 5.2, it is not difficult to imagine a vole could travel from one horizon to the other without ever being exposed to the eyes of the kestrel. Hedgerows also play an important role as permanent habitat for many species, not just as a transitory corridor.

Bringing Biodiversity amongst Crops

We have looked at the benefits of trees in creating complex permanent habitats such as woodlands and at the importance of wildlife corridors. However, helping wildlife find its way to our farm is just the start of the job. A key factor in successful NPM is reducing the distance from a habitat that beneficial insects, such as pollinators and predators, need to travel in order to have a meaningful impact. Studies show that when more than 50 to 60m apart, some species that live and travel in these habitats become less effective because their area of effect can't expand.[16] So, a field surrounded by a healthy well-managed hedge can expect a beneficial impact from this hedge at least 50m into the field. Most fields are large enough, over 60m wide, that, even with a healthy hedge all the way round, there would be a large area that is beyond the beneficial range of the hedge.

This is where agroforestry as a tool for NPM comes in. Trees and their understoreys provide that final link to take habitat from the boundaries of our fields and place them right in amongst the cropping areas. As discussed earlier in this chapter, the complex nature of tree habitats means that a line of trees in amongst your crops delivers a very diverse habitat below ground, at ground level and above crops, bringing all the different kinds of beneficial wildlife – whether insect, mammal or bird – right in amongst the crops. In large fields, we can factor in this known range of beneficials to successfully design an agroforestry system that is practical, to ensure it does not compromise the efficient working of a commercial food-growing operation, but also delivers in terms of harnessing the biodiversity for NPM. When we planned our system at Abbey Home Farm, we used our understanding of the range of beneficials, combined with a calculation of the wind-reducing effect of trees, to define the distance between the tree rows in our cropping areas.

Tree Choice for NPM

Most trees are equal when it comes to biodiversity, but, as George Orwell might have put it, some are more equal than others. While almost all trees are a huge boost to biodiversity, some provide even more of a boost. When selecting trees for biodiversity, it is important to focus one's mind again on what it is about trees that attracts this biodiversity. The other factor to consider is the potential effect of climate change on species composition and wildlife populations. What thrives in your location and climate today may not be able to survive in 20 or

Table 5.1. Factors Affecting Your Choice of Trees for Biodiversity

Benefit	Things a tree might deliver
Food source	Pollen, nectar, honeydew, foliage, nuts, seeds, berries, fruit and insects living on trees
Shelter	Nesting, perching, hollow branches, hollow trunks, leaf litter and root zone
Diversity	Supporting a wider variety of animals as well as ensuring the benefits of food and shelter are provided throughout the year due to the different tree life cycles; protecting the trees from pests and disease (trees can be vulnerable monocrops, too)

30 years. Similarly, some species that you would not have considered might be the perfect choice in a future climate scenario. We have grown almond (*Prunus amygdalus*) trees, for instance, at Eastbrook Farm. Not an obvious choice for our heavy clay in the middle of England, you might think, and so far you would be right. They have yet to produce a significant crop (after seven years), despite growing vigorously and having plenty of blossom. However, as the risk of spring frosts potentially decreases, they may yet be a successful species for us.

In the appendix, 'Tree Directory', the individual biodiversity benefits of different species of tree are described.

Native Trees

The arguments for favouring either native (to support existing biodiversity) or non-native (to adapt to a changing climate) are complex. Our suggested approach is to plant mostly native species, but to include some non-natives to provide some resilience to the changing climate. In the UK many fruit and nut species are not native but are still a good option for horticultural agroforestry systems, providing both biodiversity benefits and a cash crop. It does not matter how good a tree is for biodiversity. If it is not suitable for the soil type and the intended location, it will not thrive and therefore will not become a great habitat.

CHAPTER SIX

Getting Value from Your Trees

Calculating the financial benefit of trees in complex polycultural systems is usually very tricky due to the effects they have on a range of other factors. Reduced pesticide costs, longer growing season and increased pollination, for instance, should all help the bottom line, but knowing by how much will likely be impossible to work out. Beyond these benefits to your growing system, most trees will also produce a product that has a financial value in itself. If you are a commercial grower, this may even be the primary factor in deciding which trees to plant. Trying to work out the best way to present this chapter proved difficult. Many trees offer more than one product and potentially different markets that you could supply, so we have stuck to broad groups and provided more detail on individual trees in the appendix, 'Tree Directory'.

The first decision you need to make is how much time and effort you want to put into harvesting, processing and marketing. For instance, at Eastbrook Farm, we grow sea buckthorn (*Hippophae rhamnoides*) and supply the fruit to a gin company, which makes our branded organic gin. This is only possible because Helen Browning Organics, the business that runs that farm, also has a food processing arm and is well practised in developing and marketing value-added products. If you just want to sell fruit from the farm gate, you might grow something less fiddly and more robust, like apples or plums.

Similarly, getting value from the wood itself can be simple or complex. Using it on your farm as woodchip for mulch, compost or livestock bedding will have a relatively low value in terms of reduced input costs or improved performance of crops and animals, but on the other hand it is very cheap and easy to implement, and will use almost any type of tree. Growing top-quality timber can be lucrative but has a long return on investment (ROI) and requires careful species choice and pruning to ensure the timber reaches its top potential value. Obviously, there is no right or wrong in this. You must decide what suits your temperament, skills and potential markets.

Fruits and Nuts

Growing fruits and nuts can work very well in a horticultural system, especially if you sell direct or have another existing market for your vegetable crops that will take additional crop lines. With careful selection of species and varieties, you can spread harvest time from June through to November (and longer if you have cold storage facilities).

If space and labour are unlimited, or you are growing just for personal consumption, then you could grow a wide variety of trees. In most systems, though, you will need to make a difficult choice between the many fruit- and nut-crop trees available. You will need enough of any one crop at a time to make picking and selling it worthwhile. You might even need to invest in some harvesting equipment or pay specialist contractors to pick. These require a certain scale to make the investment worthwhile and may have additional implications for your system as a whole.

Available space will have an impact on what size trees you grow, too. Nuts generally will grow into bigger trees (with the exception of hazelnuts), while many fruit species can be grown on dwarfing rootstocks or pruned to control size.

Here are a couple of examples of systems we have used at either end of the scale. We planted a small agroforestry site with trained apple and pear trees on a 2m-tall (7 ft) wired system. The rows were 8m (26 ft) apart. The aim was to be able to pick everything by hand and sell high-quality dessert fruit to a very local market. Labour and implementation costs were relatively high, with posts and wires and twice-a-year pruning requirement. At the other end of the scale, at Eastbrook Farm we are growing perry pears, which have small, hard fruits used for making an alcoholic drink called perry. The tree might grow as tall as 15m (49 ft) and will be harvested by a contractor with

Table 6.1. Fruit and Nut Harvest Season

Tree		**Harvest month**							
Common Name	**Genus**	**April**	**May**	**June**	**July**	**Aug**	**Sept**	**Oct**	**Nov**
Rhubarb	*Rheum*	■	■	■	■				
Gooseberries	*Ribes*			■	■				
Blackcurrant	*Ribes*			■	■	■			
Redcurrant	*Ribes*			■	■	■			
Strawberries	*Fragaria*			■	■	■	■		
Raspberries	*Rubus*			■	■	■	■	■	
Saskatoon	*Amelanchier*				■				
Apricot	*Prunus*				■	■			
Cherry plum	*Prunus*				■	■			
Cherry	*Prunus*				■	■			
Mulberry	*Morus*				■	■			
Peach	*Prunus*				■	■	■		
Plum	*Prunus*				■	■	■		
Nectarine	*Prunus*					■			
Damson	*Prunus*					■	■		
Mirabelle	*Prunus*					■	■		
Greengage	*Prunus*					■	■		
Pear	*Pyrus*					■	■	■	
Grapes	*Vitis*					■	■	■	
Sea buckthorn	*Hippophae*					■	■	■	
Asian pear	*Pyrus*					■	■	■	
Apple	*Malus*					■	■	■	■
Almond	*Prunus*						■		
Hazel	*Corylus*						■	■	
Quince	*Cydonia*						■	■	
Chestnut	*Castanea*						■	■	
Pecan	*Carya*						■	■	
Autumn olive	*Elaeagnus*						■	■	
Korean stone pine	*Pinus*						■	■	
Heartnut	*Juglans*							■	
Walnut	*Juglans*							■	
Medlar	*Mespilus*							■	

machinery that sweeps the fruit from the ground after shaking. The trees are vigorous and hardy and require little in the way of pruning once established. In our grazing-based system, there are some management challenges because the mechanical harvesting requires a clean understorey of short grass, which means leaving a wide enough margin of mown pasture and not grazing livestock underneath for a couple of months pre harvest. This aside, the system is simple and cheap to operate, but demands sufficient volume of crop to justify the contractor coming out. A smaller fruit tree picked by hand might give more flexibility in cropping. Both are viable options but have different space, labour and machinery needs.

Short Rotation Coppice

Coppicing is an ancient technique for managing trees and growing particular forms of timber. It involves cutting down trees to a stool at ground level on a regular basis. It keeps the tree in a juvenile, high-growth stage and can be very productive. Not all trees respond to this method, and the frequency of coppicing will vary depending on species and the use to which you want to put the timber.

Historically oak (*Quercus* spp) was coppiced on a very long cycle (60 years) to produce wood for shipbuilding, while sweet chestnut (*Castanea sativa*) stakes can be grown on 15-to-20-year cycles. For many horticultural systems, a shorter rotation of quick-growing species, like willow (*Salix* spp), poplar (*Populus* spp), alder (*Alnus* spp) or hazel (*Corylus* spp), can offer a flexible option. Typically grown on a 3-to-7-year cycle, these species don't grow very big and so can fit within a smaller growing situation. They offer great wind protection and biodiversity benefits, and when harvested can be used either as poles for supporting crops or for chipping. There are even opportunities for growing some edible crops in this way, for instance apples or hazelnuts.

Though not yet fully explored, we are interested in furthering the work of the pioneering fruit breeder Hugh Ermen, who grew named apple varieties on his own rootstocks. Unlike normal fruit-grafting techniques or growing trees from seed (which produces trees of a new variety), Hugh rooted cuttings, which he then grew ungrafted. He believed that ungrafted trees were stronger and healthier, got a better

Table 6.2. Estimated Yield of Woodchip from Sample Tree Species

Species	Yield (ODT/ha)	Stems/ha
Willow	4–30	15,000
Hazel	2	1,500
Poplar	4–20	9,000
Alder	16	15,000
Eucalyptus	16–22	1,500
Softwoods	7–20	2,500

fruit set and produced fruit of a higher quality that kept for longer. Growing these apple varieties could work very well in a coppiced agroforestry system. Pollarding is a similar technique where you cut to a point higher on the trunk rather than at ground level, and can be similarly incorporated into growing systems.

For growers with livestock on their farm, the young growth of willow (*Salix* spp), poplar (*Populus* spp) and mulberry (*Morus* spp) provides a superb feed with high levels of micronutrients and tannins.

While there are many factors affecting yield, table 6.2 lists some guide yields if you are looking to produce woodchip from your coppiced trees.

Note that some of this data comes from high-yield hybrid willow and poplar species bred for biomass production, which might not be the best choice for an agroforestry system with vegetables. While these figures will give a guide as to how many plants you might need to produce a tonne of woodchip, they should not be used as an accurate predictor of yield. Single or double rows of trees will also perform differently to those grown in a dense coppice or woodland. The reduced competition for light and moisture could help them grow more quickly; however, they will be more exposed, which could have a detrimental effect.

Timber

Growing trees for timber is a long-term endeavour and will mostly be less suitable for smaller horticultural systems. There are ways that larger trees can be incorporated, though perhaps in areas further away from the cropping areas or if you are looking at a transitional system where the agroforestry will eventually become a woodland. Unlike short-term crops, it is not easy to grow trees for a market, when harvest and sale will be many decades in the future. For those of us in the second half of our lives, we are unlikely to witness the felling of even short-term timber species. The best we can do is choose those that will deliver other benefits in the shorter term and ensure that we manage the trees in such a way that they produce a high-quality product when the time comes. Protection from deer and squirrels is key. In the UK, squirrel damage has been estimated to cause £37 million of damage to trees per year.[1] At worst, their bark stripping will kill the tree, but even when the tree survives there is often damage to the quality of the timber.

Here are a few small or very fast-growing species that might be worth considering for a vegetable growing system.

Dogwood (*Cornus* spp) and wild service tree (*Sorbus torminalis*). Both are relatively small trees that do well in high-light situations. With specialist markets in musical instruments and woodturning, it may be possible to get a relatively good price for a small number of logs.

Cricket-bat willow (*Salix alba* 'Caerulea'). While most willows will not provide valuable timber, there is a good market for cricket-bat willow. But you will need very wet conditions for it to grow to the grade required.

Paulownia (*Paulownia tomentosa*). This is very quick growing and late to leaf, so it is suited to vegetable systems where early-season light is important. The large leaves can also provide summer shade if required. Paulownia foliage is extremely high in nitrogen and coppiceable. The timber is very light and strong.

Birch (*Betula* spp). As a pioneer species, birch grows quickly and has a relatively upright growth and fine-leaved canopy so doesn't cast heavy shade. However, some birch species, for instance silver birch (*Betula pendula*), has shallow wide roots that can compete with underplanting.

Cherry (*Prunus* spp). Although it's not usually possible to combine a fruiting cherry crop with timber output, cherry trees are not huge, are superb for wildlife and can be harvested for timber after 30–40 years.

Bush Fruit and Soft Fruit

Growing shorter woody species in a horticultural agroforestry system can work well. They tend to fruit more quickly than tree species, so provide a quick return while we wait for our larger trees to start producing a crop. Planting between larger trees can use the space that would otherwise be empty until that tree reaches its final size. Alternatively, rows of bush fruit can be planted alongside the trees to protect and encourage upward growth. Picking bush fruit is very time consuming and most have a short shelf life. On-site refrigeration or freezing facilities can help prolong produce life and reduce the risk from gluts that have no immediate market. For those growers interested in adding value, most bush fruit processes well into a range of products.

Our experience at Eastbrook Farm suggests that it does matter what you plant under or between the trees, though we do not have a definitive list of recommendations yet. Trees with dense canopies and spreading habits will tend to outcompete the bushes planted between them, while trees such as birch and almond with more upright habits and smaller leaves are better suited to interplanting. Trees that produce suckers, such as most *Prunus* species, may also cause problems. The suckers can grow up through the bushes and can be almost impossible to remove.

Herbs and Floristry

Though a relatively underexplored area of agroforestry, we believe there is huge potential for marketing herbs and flowers, both from trees themselves and from under- and inter- planted crops.

Herbs

The global herb market is worth many billions, and in the UK, for example, we produce almost none of that ourselves. There are practical reasons for this, including price of labour and a damp climate that makes drying more expensive. However, within an integrated system where the plants are providing multiple functions, the margins on the saleable crop can afford to be smaller and, therefore, the potential for adding income from herb products is worth considering. As with nuts, there is work to be done on building processing capacity and collaboration with other growers is likely to be key.

There are a number of trees with herbal potential from almost every part of the plant. Flowers from lime or linden (*Tilia* spp) and witch hazel (*Hamamelis* spp), leaves of ginkgo (*Ginkgo biloba*) or sea buckthorn (*Hippophae rhamnoides*), bark from trees such as willow (*Salix* spp) or birch (*Betula* spp), or seeds from the ash (*Fraxinus* spp) are just a few examples. Unlike nuts, fruit and vegetables, most herbs will require almost immediate on-site processing of some sort, usually drying or, if you are ambitious, extracting and distilling the oils. For those interested in learning more about this venture, we recommend reading *The Organic Medicinal Herb Farmer* by Jeff and Melanie Carpenter.

Flowers

Flowers are easier to harvest and sell than herbs, but you need a physically close market because shelf life is an issue. Pick-your-own is becoming popular and it's a great way of getting income without high labour cost. Also, adding mixed bunches or buckets to your market stall or farm shop attracts customers and boosts income. Building relationships with florists is another big opportunity, either for specific flower crops or for foliage plants. Coppicing ornamental species such as twisted hazel (*Corylus avellana* 'Contorta') and twisted willow (*Salix matsudana* 'Tortuosa'), copper beech (*Fagus sylvatica* f. purpurea) or eucalyptus (*Eucalyptus* spp) will produce lots of saleable material. Shrubby species such as butcher's broom (*Ruscus aculeatus*) and photinia (*Photinia* × *fraseri*) are also perfect for interplanting amongst bigger trees. You can also incorporate perennial, and even annual or biennial, flower species into the understorey of your trees. In the early stages of your system, when the trees are small, this is a great way of getting good income from the row gaps. However, there are still plenty of species that can tolerate semi-shade and produce useful material once your trees are larger. Table 6.3 shows a few examples, but it is by no means exhaustive.

Table 6.3. Trees and Plants for Floristry

Plant type	Examples
Grasses	greater quaking grass (*Briza maxima*) northern sea oats (*Chasmanthium latifolium*) hare's tail grass (*Lagurus ovatus*)
Annual/biennial flower that grows under trees	sweet William (*Dianthus barbatus*) foxgloves (*Digitalis purpurea*) bulbs (*Narcissus* spp, *Allium* spp)
Perennial flowers	*Sedum* spp lady's mantle (*Alchemilla mollis*) *Achillea* spp *Paeonia* spp *Echinops* spp
Shrubs	*Viburnum* spp *Spirea* spp lilac (*Syringa* spp) *Buddleja* spp *Ruscus* spp *Photinia* spp
Twisted tree forms	hazel (*Corylus avellana* 'Contorta') willow (*Salix matsudana* 'Tortuosa') dragon mulberry (*Morus alba* 'Unryu') beech (*Fagus sylvatica* 'Tortuosa') dragon's claw willow (*Salix babylonica f. tortuosa*)
Winter colours	willow, including *Salix alba, S. daphnoides, Salix x rubens Basfordiana, S. fargesii* *Cornus sanguinea*
Foliage	*Eucalyptus cinerea* hazel (*Corylus avellana*) oak (*Quercus* spp) holly (*Ilex aquifolium*) beech (*Fagus* spp)

Future Crops from Trees

We come from a vegetable-growing background where you can judge success or failure within one season. If a crop doesn't work, we can do it differently next year; we might try a different variety, plant earlier or later, weed it better or carry out any number of other potential tweaks to our growing methods. We can also respond quickly to market trends. Kale suddenly becomes fashionable. That's great, let's plant a bunch of it to meet that demand.

Trees are rather different. Depending on which part of the tree we are selling, they can take a long time to mature and reach cropping age. There is a degree of crystal ball gazing involved. At Eastbrook Farm, when we started planning our tree planting, we worked out a 25-year business plan. We knew full well it was not likely to work out exactly as planned – for instance, no one was growing organic almonds in the UK, so we could not accurately predict yield or market price (or even if they would grow in our soil and climate). However, we were able to roughly estimate when trees were likely to crop and therefore when we might expect income from the different types of trees.

There is such a wide potential of products from trees that looking into the future and keeping options open for new markets is worth the effort. Here's a few areas that are taking off, and, for those growing on a larger scale, it might be worth investigating. Some of these ideas will require specific tree species or timber qualities, while others might be flexible and able to use lower-grade mixed woodchip species.

Microfibrillated cellulose. This is an area of intense development as companies look to remove plastic packaging from their supply chains. Typically, the market is looking for fast-growing pulp species, such as poplar (*Populus* spp) and spruce (*Picea* spp), but they are increasingly looking to a wider range of species as demand for wood pulp increases. One study looked at the opportunity for using olive prunings, for instance.[2] So as a byproduct of agroforestry coppiced or pruned systems, this could have potential. We especially recommend looking for small local innovative businesses that might be looking for smaller quantities of a particular material.

Cross-laminated timber. Conifer species, such as spruce (*Picea* spp), pine (*Pinus* spp) and larch (*Larix* spp), are probably less suitable in a horticultural setting, but if grown as part of a mixed windbreak, the product could end up going on to produce cross-laminated timber, which is increasingly being used as an alternative to concrete or bricks to form the walls, roofs, floors and ceilings of a building, especially in multistorey construction.

Biodegradable electronics. There is even research into the use of wood cellulose for making biodegradable electronics.[3] Again, though this might not be a market we all want to tap into, it shows the breadth of potential uses for trees.

Routes to Market

It is a cliche, but for a reason; always grow for a market! This is easier for annual or quick-growing crops, but even with trees it helps to try to define the market. In the early planning stages, we find it helpful to break down the potential markets as follows.

Internal or 'On-Farm' Markets

These are ways of using the produce on your own farm – for instance, coppiced stakes for fencing, woodchip for mulching or willow browse for livestock. This is a guaranteed market with almost no transport cost and no margin lost to middlemen. Being able to identify which external inputs you might be able to replace with home-grown alternatives is a great starting point when setting your objectives. In stock-free systems, which many horticultural holdings are, building and maintaining fertility can be a major challenge. Using trees, and especially nitrogen-fixing species, is one way to reduce reliance on bought-in nutrients.

Local Sales

This market might include your own sales through a box scheme, or a farm shop in the case of fruit, but it could be wider if you plan to grow non-food crops. Flowers to your local florists, bean poles to a garden centre or fence posts and woodchip to other farmers are just a few of the opportunities. This approach lends itself to smaller growers with a mix of products.

Wholesale

If you would rather grow larger quantities of one or two products, then finding wholesale markets is likely to be a better option. Though margins are likely to be smaller, efficiencies of production and distribution can make this more profitable, particularly for what might be a 'side enterprise' fitting alongside your main vegetable growing business.

Added-Value

Adding value to your products can be profitable, even though it adds complexity and cost. Almost any product from your trees could be further processed to add value: fruit jams, nut butters, propagation substrate from your composted woodchip, turned bowls from the timber. Working in partnership with family members or bringing in other collaborators is another way to build on your primary production.

Contract Growing

Growing for someone else reduces your risk and helps guarantee a sale at the end of production. Cricket-bat willow (*Salix alba* 'Caerulea') is often grown in this way, with specific companies providing trees and advice. Apples for cider is another UK example of historically long-term contracts that have been offered to growers to secure supply for the company. The risk here is that you may end up being tied into less favourable contracts or risk paying back investment money if crops fail.

Contracting Opportunities

Though it might not be the best option when you are starting out, once you have some skills, experience and possibly specialist machinery, there could well be opportunities to hire yourself out to other growers: harvesting woodchip, for instance, or drying herbs or perhaps helping to look after trees for other growers.

Sourcing Trees

If you haven't had to buy trees before, the choice and requirements can be a bit bewildering. With the recent explosion of planting, getting hold of some types is also becoming tricky. We won't go into specific suppliers but here's some general guidelines to help you navigate.

Broadly speaking, there are three options when buying trees: bare-root, cell-grown and pot-grown. This is further complicated by the difference between buying trees for forestry or woodland planting versus specialist fruit and nut species versus particular ornamental forms for floristry.

Bare-Root Trees

These are trees grown in a field and then dug up as soon as they become dormant and sent to you to be planted before the following spring. This is the cheapest way to produce trees, and in my experience mostly gives the best results in terms of establishment and growth. Since the roots are often smaller than pot-grown trees, planting also tends to be cheaper and quicker.

However, there are risks. Unless you intend to plant them as soon as they arrive, you will need to 'heel them in'. In other words, put them in the soil, compost or woodchip to protect the roots from drying out until you can plant. If they are not heeled in properly, or if voles, mice or other pests nibble them while in storage, they can suffer. The other challenge with bare-root trees: as the changing climate sees winters both warm and shorten in many parts of the world, the planting window for bare-root trees is getting shorter. I have had trees being lifted while still in leaf, as the nurseries would not otherwise be able to physically lift all their trees in time. In addition, professional tree planters will be very busy during the short planting season, which is a problem if you don't have the labour to plant your own.

Most common species are sold by height, for instance, '40–60cm'. Smaller trees are cheaper and quicker to plant but are a little more vulnerable in their first year or two. Some planting grants may require you to buy a particular specification, so look out for that before ordering.

Cell-Grown Trees

Using techniques well understood by vegetable growers, these are like plug plants, or modules. Some companies grow exclusively by this method, and some tree species, especially conifer species, do better when cell grown. They are more robust, less likely to be damaged during transit and, importantly, can be planted both earlier and later than bare-root trees. However, they are generally more expensive to buy and transport since you have all the compost as well as the tree itself. When it comes to planting, they are also more cumbersome to move and carry when planting. This might not be a big issue if you are planting a few hundred, but once you get up to thousands and are employing professional planters this can make quite a difference. We have found that they don't always take as quickly as bare-root, though they do often then catch up as they mature.

Pot-Grown Trees

This method is less common for woodland and timber trees, though some species, like holly (*Ilex* spp), do better in pots than bare-root. Pots are most commonly used for specialist ornamental plants or grafted fruit and nut trees. As with the cell-grown trees, pot-grown gives more flexibility on planting times, but the plants are more expensive and can be a problem to establish. You will also need to dig a bigger planting hole to incorporate the root ball, so planting times can quickly escalate.

Some Other Thoughts on Tree Sourcing

Organic

We both work on certified organic holdings and so are required to source organic trees where possible. Historically in the UK there were only two nurseries selling organic fruit trees; this has now increased to five nurseries and a wider range of species. However, there are still many species we cannot get organically, so we look for the best supplier of trees.

Local Provenance

If you have a very local nursery or can raise your own trees (see the next section, 'Grow Your Own Trees'), your trees are more acclimated to your local conditions. For trees grown from seed, you may also want locally sourced genetics. Worth noting though that with climate change you may also want to hedge your bets by including some genetics from climates that match the prediction for your locality in 50 or 80 years' time. Increasingly, we are also becoming aware of the risk of importing diseases and pests with trees. Ensuring that your suppliers comply with tree health requirements is also essential to help reduce this risk.

Grow Your Own Trees

With increasing costs, uncertain availability and risk of introducing pests and diseases, there is a burgeoning interest in growers for producing their own trees. As growers, we are mostly well versed in propagating from seeds and cuttings, so applying that vegetable knowledge to trees is not a big step. We've done some tree propagation and would encourage growers to have a go but, as you would expect, some species are easier than others.

Willow (*Salix* spp) is the most straightforward. My old college lecturer used to say, 'That'll root in your pocket, that will.' Using water that has had willow soaking in it will even help other cuttings to root, such is the strength of the rooting hormones contained in the willow wood. Almost any length of willow cutting will root if you stick it in the ground. We are experimenting at Eastbrook Farm with 3m-long branches driven into the ground to give enough height to grow away from the many deer we have roaming the farm. Poplar (*Populus* spp) and common dogwood (*Cornus* spp) will also grow from hardwood cuttings. However, bear in mind that any vegetative propagated trees will be genetic clones, and therefore a system relying entirely on these will be potentially less resilient.

Growing trees from seeds will provide wider genetic diversity, which may give more ability for some trees to respond to changing environmental challenges. Some trees grow easily from seed, like oak (*Quercus* spp), cherry (*Prunus* spp) and apple (*Malus* spp) (though bear in mind that apple may not produce good edible fruit, instead being useful for wood and other products). Other species can be very tricky, needing specific stratification – in other words you must replicate the conditions that the seed would go through in the wild, typically a period of cold. You can collect your own tree seeds (though do make sure you have the landowner's permission) or buy in seed, either pre- or post-stratified. Unlike propagating commercial vegetable seeds, achieving anything over 90 per cent germination is tricky, so factor that in when calculating numbers.

As with vegetable crops, you will need to keep your tree seedlings watered, weeded and free from pests. Rabbits and voles are particularly keen on tree seedlings. You can set aside an area separate from your vegetable garden, or incorporate tree production as part of your rotation. Controlling weeds is often easier within the rotation than in a separate area, though if you are using mechanical weed-control methods the trees may grow too tall for this to work, especially if you grow any on for a second year.

Grafting is a method used particularly for known varieties that don't take from cuttings, or where you want a tree of known vigour. Grafting is relatively straightforward and doing your own can significantly reduce the cost of fruit and nut trees. Beware that most of the modern varieties are likely to be subject to plant breeders' rights, so permission is needed before propagating.

CASE STUDY
Nathan Richards and Alicia Miller, Troed-y-Rhiw

Troed y Rhiw is a 23-acre mixed organic farm situated in West Wales on Ceredigion's coastal belt, less than 2 miles from the sea. Established in 2008, it's owned and run by Nathan Richards and Alicia Miller. There are around 15 acres in horticultural production at any point in time, including 12 acres in active fruit and vegetable production and 3 acres in fertility building. They also have a small Highland cattle herd. Being close to the coast, it's very windy, so they maintain high and wide hedges and small fields. Their largest field is just under 5 acres.

Diversity is the key for Nathan and Alicia as they build a greater ecology of the whole farm, feeding human beings and still being financially viable. They produce only year-round fruit and vegetables for sale in their 120-customer-a-week box scheme, and a weekly farmers market, describing themselves as a hyper local, seasonal business.

Nathan and Alicia were primarily looking to increase the range of their crops for sale into their farmers market stall. They also wanted to maximise the light and heat captured by their tunnels. These are a valuable resource and they saw that growing mostly low crops was wasting potential. There is also some outdoor agroforestry with annual vegetable crops grown between rows of bush fruit.

Nathan describes himself as a 'climate change farmer', and this led them to think about their most valuable growing space, the polytunnels, and to look at how to better use the time and space within them as the climate changes. In a rotational tunnel system, there are many years with low crops. Only tomatoes and cucumbers are tall, so space is wasted at the top. People will still want to eat fruit, but the practicality and cost of importing from around the world might become more challenging. Nathan has a long-term view and is planning for opportunities that might be created not just in his own lifetime but decades into the future.

OUTDOOR AGROFORESTRY

They have a top-fruit orchard and a bush-fruit agroforestry system consisting of raspberries and currants in rows. The spacing between rows varies, with the narrowest being 3.5m (11 ft) apart while the wider ones are 12 to 15m (39–49 ft) apart. Vegetables are grown on 1.8m (6 ft) beds in between the rows. After an unsuccessful trial of heritage wheat between the bushes, they now grow courgettes, sweetcorn and beans in five beds between each row, on a 7-to-8-year rotation with 2 years of green manure crops followed by 5 years of vegetable cropping. The vegetable beds are cultivated with the 90hp tractor, the same machinery used for the rest of the farm, which keeps it simple.

TUNNEL AGROFORESTRY

There are nine large tunnels at Troed y Rhiw, mostly 9m wide × 30m (6 × 98 ft) long. Realising that cherries and other stone fruit would not be viable outside due to both the weather and high bird populations, they planted stone fruit in single rows into some of the tunnels in 2018/19. They are fan-trained to keep them narrow and prevent too much shading on the other crops.

KEY OUTCOMES

The trees produce well, and the fruit sells very quickly at farmers markets. There do not appear to have been any negative effects, either from shading or nutrient loss on the other crops grown in the tunnels, which is crucial to making the finances work. When we chatted to Nathan about his system, he made a point that illustrates how the benefits of diverse systems are sometimes hard to quantify: 'If I didn't have the trees in one tunnel, I could grow an extra 50 lettuces. If I turned up at the market with 50 more lettuce, no one would notice, but when I arrive with freshly harvested cherries and peaches, everyone comes to my stall excited to see, try and talk about them. While they are there, they buy more of my other stuff.' So, whether the cherries in themselves are hugely profitable or not, they add significant value to the business.

LESSONS LEARNT

Nathan is keen to emphasise that he doesn't consider himself an expert. 'The book for this doesn't exist for growers in the UK, especially in West Wales and in a changing climate. We are still feeling our way.'

Pruning is one of the biggest challenges. You wouldn't do any winter pruning of apricots and nectarines because of disease risk, but it's hard to know exactly when to do it. Nathan tends to prune at the first inception of blossom in mid to late Feb. However, this year they had to winter prune as the apricots were growing into the plastic at the top of the tunnel and would have caused damage. 'If I was starting again, I would have been more radical about cutting the leader out at the start, to encourage strong laterals, which would then send leaders up to create the framework. We even took out one apricot altogether one year. It had initially been planted to improve pollination, but it was on dwarfing rootstock and just didn't perform.' The other apricots are on St Julien A rootstock, and they haven't had any suckering.

Fruit selection and thinning is something they are also looking at. Although Nathan finds it hard to bring himself to thin, he would like to aim for slightly larger fruit. Having said that, they have no trouble selling the smaller fruit at the farmers market, though this obviously might be a problem for growers trying to access more traditional wholesale markets.

They have found a way to keep birds off their precious fruit by sealing the ends, either with the whole door, or by replacing the top half of the door with a wide bird netting that keeps out birds but allows access to predators and pollinators. This also gives him flexibility when trying to keep the temperature balanced to meet the needs of both his vegetable and fruiting crops.

Outside they have mostly given up on the raspberries, which Nathan says basically just feed the blackbirds. They have also moved their gooseberries from the field, where they suffered heavy losses from sawfly. They now interplant their stone fruit in the tunnels with gooseberries.

Looking at it purely financially, they might just do cherries. The product is superb quality, relatively easy, and high yielding. Peaches and nectarines do OK, and Nathan feels they can improve their system for these. Apricots are the trickiest – and particularly the challenge of good pollination with their very early flowering. This year, they might try to hand pollinate or to bring bees into the tunnel. However, honey bees don't do well in tunnels and can get caught on top of them, which is not ideal either.

Because watering is key, they use a mains-fed base-sprayer line, which allows them to give the fruit more water than the vegetable crops, which is crucial at some points in the year. They also water by hand if needed.

At the moment, economics don't add up to do more trees commercially. The cost of tunnels is high and not optimal for larger-scale fruit production, but in 20 years' time it might compare favourably to importing from Spain. At Troed y Rhiw, they might just have found a way to grow consistent-quality organic stone fruit in challenging conditions year on year.

CHAPTER SEVEN

Designing the Layout

Congratulations, you are now at the exciting and creative part of the design process. Having thoroughly assessed the proposed site, identified potential opportunities and challenges, appreciated the complex interactions, both positive and negative, between vegetable crops and trees, and selected a variety of desired tree species appropriate to your site and system, you can finally put all this knowledge together into a cohesive and dynamic plan.

Layout Options

There are almost infinite variations of layouts for a particular site, but we can group them together into a few types. There are some more suited to livestock-based systems that we won't cover here in detail. Although most vegetable agroforestry systems are based around alley cropping systems, there are other options and layout considerations worth mentioning, especially if one is adopting a whole-farm approach. Again, we urge readers to be inspired by other people's designs but resist copying exactly. Your site deserves a unique layout.

Alley Cropping

At all scales of vegetable production, there is a tendency to grow plants and crops in straight lines, as it is more efficient. Straight lines facilitate faster and more effective planting, weeding, watering, harvesting and, crucially, mechanisation. For this reason, many gardeners and growers – almost all at a commercial scale – use linear systems based around bed systems. Accordingly, for the majority of vegetable growers looking to integrate trees, an alley cropping system is often the most logical choice. This system is based around a single or multiple parallel rows

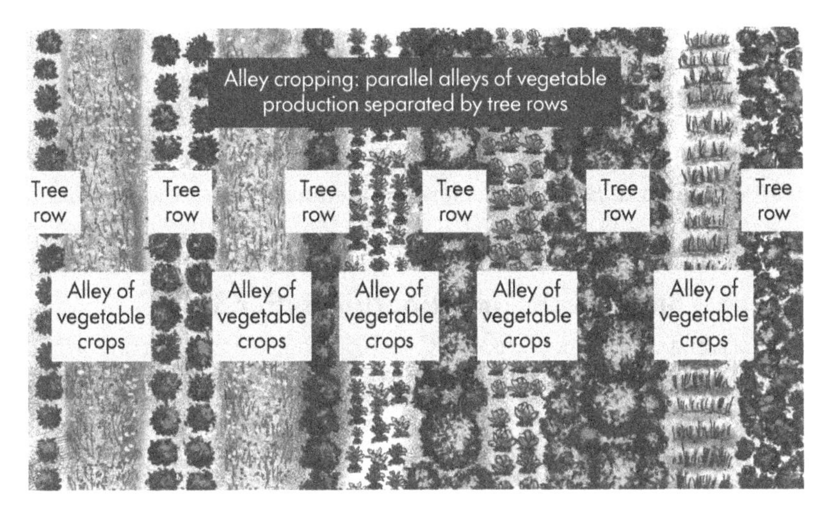

Figure 7.1. Alley cropping.

of trees with alleys of crops grown between them. The width of tree rows and alleys is up to you, and we will cover this later in the chapter.

Alley cropping designs facilitate effective management of sunlight, root-pruning strategies, and weed control of tree rows and vegetable crops. They can also help with efficient irrigation and accurate tool and machinery usage, as well as maximising effective use of crop protection meshes. Although an instinctively attractive layout, it is important to appreciate the limitations that aspect and topography can impose on alley cropping, which is discussed later in the chapter. It is also worth noting that tree lines do not have to be straight. Most of the benefits of alley systems will be achieved with curved or wavy lines and alleys, they are just a bit trickier to mark out and manage.

Forest Garden

Also known as a food forest, this is a system based on trees, shrubs and perennial plants planted in such a way as to mimic the structure of a natural forest. Although forest gardens are highly productive and used throughout the world, they tend to be most suited to home gardening or subsistence farming and are rarely, especially in temperate climates, a compelling way of producing food commercially. Although effective at producing abundant calorific edible crops, they often struggle to produce meaningful amounts and desired varieties of vegetables for the current consumer. Although it is a system well worth considering in some situations, there are many good, detailed books on this subject already, to which we would direct the interested reader. These include *The Home-Scale Forest Garden* by Dani Baker, *Creating a Forest Garden* by Martin Crawford, *Edible Forest Gardens*, Volumes 1 and 2, by Dave Jacke and Eric Toensmeier.

Hedges, Shelterbelts and Riparian Buffer Strips

As described earlier, trees are a powerful tool for managing strong winds and soil erosion, and they would contribute this benefit throughout your cropping areas. However, it is worth considering more extensive and denser plantings around the edges of fields or farms and alongside watercourses. Think, too, about how these features link together, both within your own holding and with the wider landscape.

Block Plantings

These are groups of trees, often in corners of fields, patches between buildings, or any other chunk of land into which you don't plan to plant vegetables directly. They are usually of less benefit in purely horticultural settings from a shelter-and-shade perspective but are potentially very useful for biodiversity or materials like woodchip or stakes. These types of planting also suit those with livestock, providing a cost-effective way of establishing dense groups of trees for the benefit of animals.

Retro Planting

While we think mostly of planting new trees when looking at horticultural agroforestry, you might want to incorporate vegetables into land that already has trees. It's not an option that is often covered in agroforestry design. However, vegetable growing is the most productive food group per metre squared, making it possible to squeeze vegetable production into the most unlikely places. Integrating vegetables underneath existing trees should not automatically be discounted. However, it is particularly important, if you decide to plant under old, established trees, to pay particular attention to the location of the roots and plan how to manage their interaction with your vegetables, particularly in terms of irrigation.

However, there are many derelict orchards in both cities and the countryside that could be used in this way. Perhaps one of the most visually pleasing images of vegetable agroforestry is a system established by Moreno de Meijere at Aromath Farm in France, where he and his family planted a market garden under an old, established orchard.

Tree Density and Spacing

Tree-planting density is determined by your objectives and chosen tree species. For instance, to get long, straight, quality products from your trees (either from timber or coppice products), you will need to create competition between your trees by planting a dense stand of either mixed or single species. It is the competition for light from close tree spacings that results in straight vertical growth. Let's take common hazel (*Corylus avellana*) as an example. Traditional plantations of coppiced products for fencing, thatching and horticulture would have trees spaced at 2.5m (8 ft) in all directions. So, if your objective is hazel bean poles and trellising material, you would not just need close spacing between trees but also a wide multiple-tree row to allow for increased density and competition, giving a straight end product from the trees. However, if you were planting hazel for nut production, then 4m (13 ft)

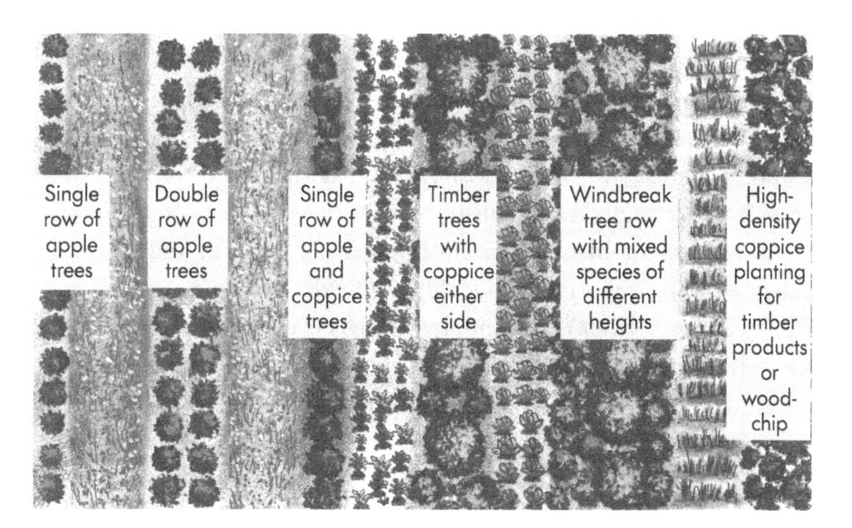

Figure 7.2. Tree row densities.

Table 7.1. Spacing Guide by Rootstocks

Rootstock vigour	Very dwarf	Dwarf	Semi dwarf	Semi vigorous	Vigorous
Rootstock apple examples	M27	M9	M26	MM106	MM111
Suggested tree spacing	1.5m (5 ft)	2.5m–3m (8–10 ft)	3.5m (11 ft)	4m (13 ft)	4.5m (15 ft)

spacing would be more appropriate and they could happily be planted in a single row using an alley cropping system. The higher the planting density, the more pressure on water and nutrient competition with surrounding vegetable crops. One way to mitigate this might be to leave a wider gap between the trees and vegetable rows.

When planting fruit trees, spacing is determined largely by the rootstock of the trees and whether you are interplanting them with other species of fruit tree. Looking at apples for instance, see Table 7.1.

Different sites and soils will impact the choice of fruiting rootstocks. The vigour of the rootstock can be used to offset inherent issues with site or soil. For example, selecting a more vigorous rootstock will help trees cope with the difficulties of growing on particularly exposed sites, shallow soils or shady positions. At Abbey Home Farm, we planted apple trees on MM106 rootstocks. In hindsight we should have used a more vigorous rootstock in the areas where we knew the soil was shallower, as the apple trees in these areas have struggled to establish.

We interplanted with common hazel (*C. avellana*) and common alder (*Alnus glutinosa*) for woodchip, and elder (*Sambucus nigra*), wild cherry (*Prunus avium*) and spindle (*Euonymus europaeus*) for biodiversity. They were planted in a single row on spacings demonstrated in Figure 7.3.

Our density is relatively low to allow easy access for harvesting of apples and woodchips. It should also allow light to reach the wildflower understorey. The main objectives of the project were to reduce wind speeds, become self-sufficient in apples, enhance predatory insect

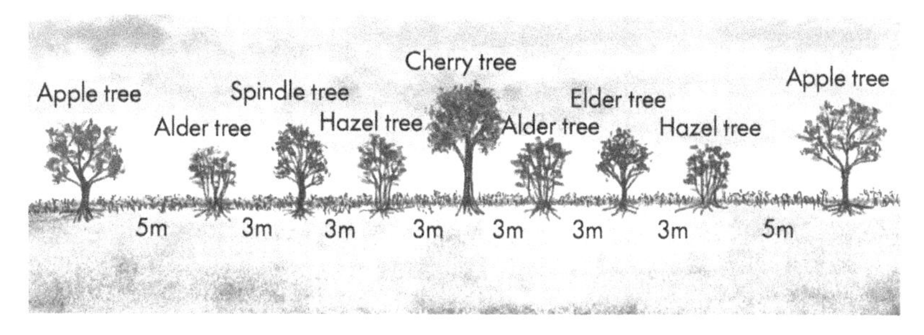

Figure 7.3. Tree spacing at Abbey Home Farm.

populations and produce woodchips for composting. High-density planting is not essential for any of these objectives.

Row Width and Length

In an alley cropping system, the width and length of a tree row has a direct impact on the area available for vegetable production. There will be a loss in the vegetable cropping area, so factor it in at this stage of planning. However, over the long term, yield per m² should be offset by additional three-dimensional cropping from the trees and the other benefits of trees to crops. This adjustment is particularly important when establishing trees into an existing system that is already delivering specific yields of vegetables for existing outlets. There is a time lag between establishing your trees and realising their products and benefits; therefore, expect an initial drop in vegetable yields and financial income, or compensate by taking on more land and expanding your overall area.

Because of this initial drop, it makes financial sense to try to achieve your objectives with as narrow a tree row as possible, particularly at the start. Once the trees are delivering, you might allow rows to become wider.

The initial tree-row width is determined by the size of trees being planted, the planting density, the growth rate of that tree species and the system it is going into. For example, at Abbey Home Farm, our vegetable production is on 1.5m-wide (5 ft) beds, so we selected this as the initial row width. This let us use existing equipment to prepare

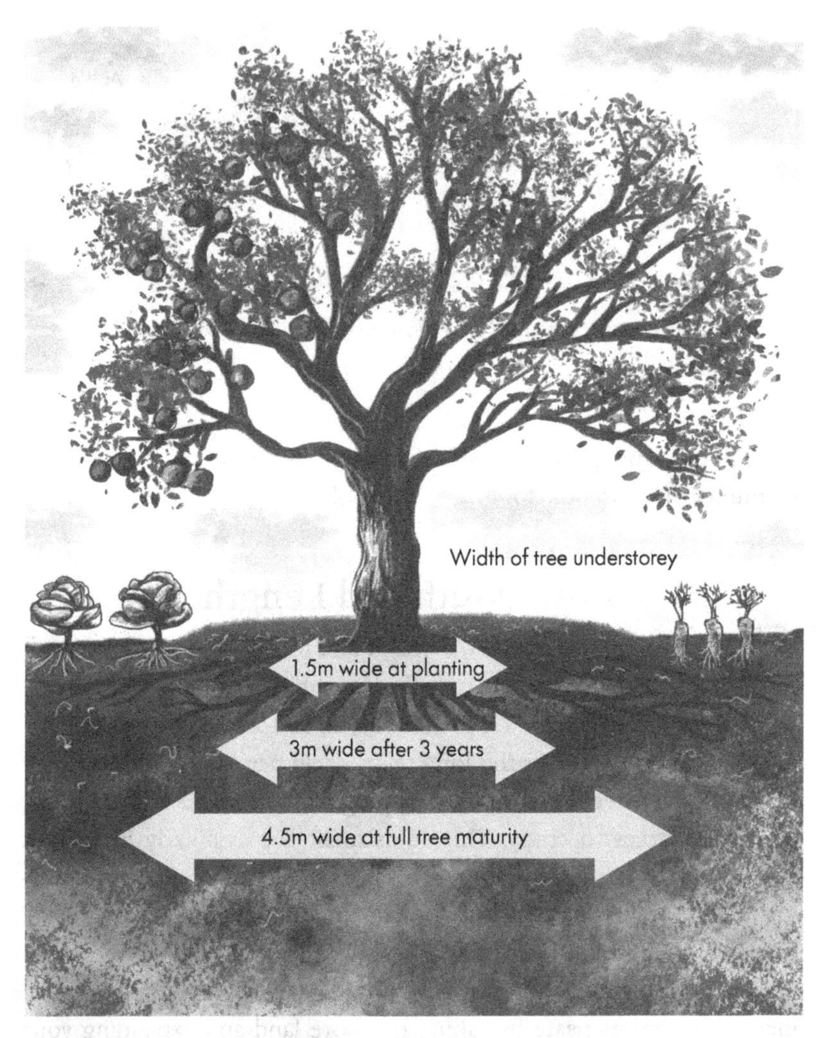

Figure 7.4. Expanding tree-row width.

rows with cultivating and weeding machinery. It also maximised our productive vegetable area while the trees were establishing.

As the trees grow, their rooting systems spread. This is normally accommodated in two ways: first, the tree row is allowed to become wider, and then, when the desired tree-row width is achieved, root pruning is used to direct roots below vegetable cropping depth. The width of the tree rows at maturity will be influenced by tree species and their habit. For fruit trees, rootstock will also shape row width. As a general rule, the width of a tree's rooting area is twice the canopy width. Fruit tree and coppice cropping systems end up with tree row widths between 2 and 4m (7–13 ft) at maturity, if planted in single rows.

The length of a tree row often matches the length of cropping areas to maximise tree crop yields and benefits. However, don't forget that you will need access around the end of tree rows to enable machinery access both from the field entrance and between the alleys themselves. Work out the turning circle of your biggest piece of current or planned machinery and use that as the basis for the width of your headlands.

Alley Width

Alley width is the distance between the tree rows, once again determined by your objectives. Nutrient recycling, for example, requires relatively narrow alleys, whereas biodiversity management and wind reduction can work with wider alleys. Some basic measurements can assist in helping with this decision, based on the effect that trees will have on wind and light.

$$\text{height of tree} \times 10 = \text{distance of effective wind reduction downwind from trees}$$

Well-designed windbreaks generally reduce wind speeds downwind of the trees at a distance ten times the height of a tree or tree row. This means a 5m (16 ft) tree will have an effect for 50m (55 yds).

$$\text{height of tree} \times 2 = \text{minimum width of alleys}$$

As a guideline, we recommend that the alleys should be at least twice the width of the height of the trees at maturity. The minimum width of alleys is usually governed by shade cast by the trees; the taller the tree, the further the shading effect. As discussed earlier in the book, even if shade is to be a key objective it is still important to allow sufficient light to the vegetable crop. We are not looking for total shade. So, with smaller trees, like fruit and coppice trees, which often don't reach much higher than 5m, one could have alleys that are

10m (33 ft) wide. But if growing taller trees, for example for timber, wider alleys will be required.

maximum distance travelled by beneficials = 50m (55 yds)

The effective range of most predatory insects from a permanent habitat is 50m. This would be a good maximum width for alleys. Other factors on the farm may dictate wider alleys, but be mindful that you would lose some of the NPM benefits.

Systems Approach

The previous calculations will give you an estimate of how wide your alleys could be, but of course you need to also make it work with the rest of your production system. Alley width needs to dovetail with the widths of your beds, irrigation kit, machinery and crop protection mesh. Normally this is done by picking the largest of all of these and then making alleys a multiple of that.

For example, at Abbey Home Farm, our alleys are all multiples of the width of our irrigation boom. The boom is 16m (52 ft) wide, which irrigates 11 beds, each with a width of 1.5m (5 ft). We have alleys that

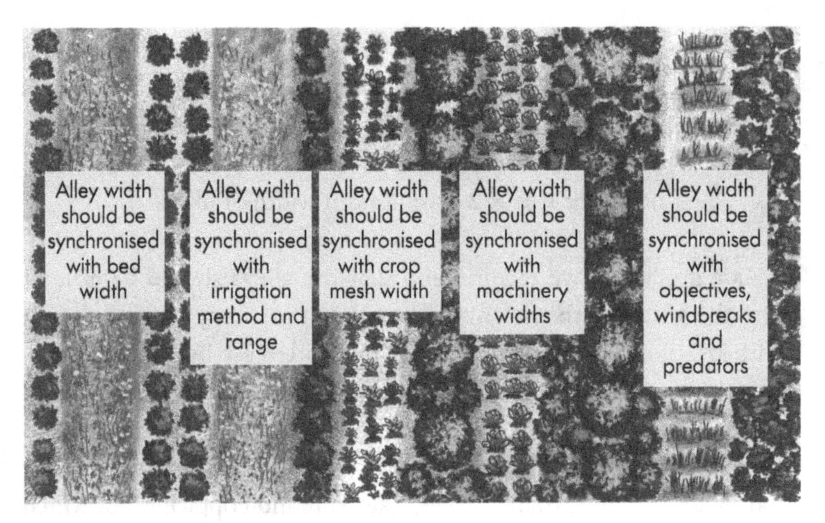

Figure 7.5. Systems approach.

are 50m (55 yds) wide, which fit three widths of the boom, with a 1m (3 ft) gap to the trees. This allows for the trees to be within range of the water from the irrigation boom, so that trees and understorey can be watered at the same time as vegetable crops. This has been particularly helpful for establishing trees during drought conditions. Our crop protection meshes are 4 beds wide, so we grow 32 beds of vegetables in the alleys, which allows for efficient use of meshes.

Aspect and Topography

When designing a system with so many complex and interacting variables, there will always be a need for balance and compromise. The two factors that are often likely to require the most compromise are aspect and topography.

The most notable of these is the orientation of tree rows within an alley cropping system. What might be perfect in relation to the sun and prevailing wind could be impossible or impractical when taking the topography of the land into account. Row orientation can be the single most influential decision in the design process. If we then also factor in the shape of the field, which might not be alterable, you might find yourself faced with some hard choices. We worked with one farmer who decided to remove an existing hedge to create a workable alley system across a larger area. While this is an extreme example, it demonstrates the potential limitations of working within existing boundaries, both natural and human-made. Once again, the objectives you have set for yourself will help guide you through this sometimes tricky part of the design process.

Aspect

Light, wind direction and slope are the three fixed considerations of aspect. Field shape and the direction of current existing cropping are also important but sometimes more flexible variables. We will look at slopes in the 'Topography' section on page 108. The others we will discuss here.

LIGHT

To capture as much of this valuable resource as possible in alley cropping systems in higher latitudes, we orientate rows in a north-to-south

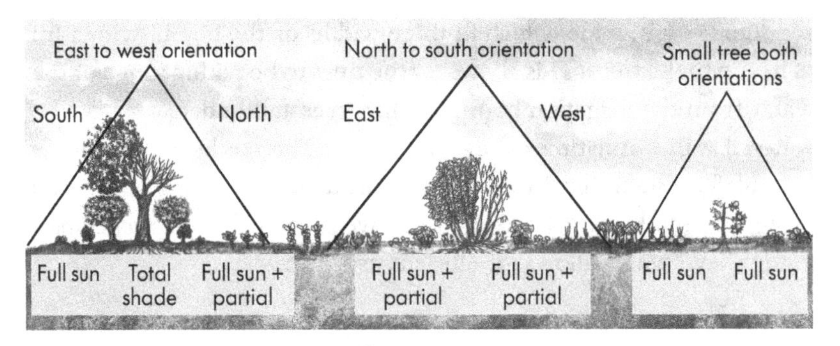

Figure 7.6. Shading effects (east–west, north–south).

direction. Although there are ways of managing the permanent areas of shade to the north of the tree rows that result from an east–west planting, it is preferable to be as close to a north–south orientation as possible. This allows the alleys on both sides of the tree row to receive equal amounts of light as the sun passes across the sky. It also allows light to reach all areas of the tree understorey, which opens up more cropping options.

However, the ideal of a north-to-south orientation is not always achievable. If your objectives demand that tree rows run in a different direction, there are ways of offsetting this. Managing tree height reduces shading on the north side of a tree row. This can be achieved through species choice or by management techniques that keep trees to a certain height, such as pollarding or coppicing. Increasing tree spacing and decreasing planting density is another way of allowing more light to pass through the tree rows. We would also avoid evergreen species if light transmission is a key objective.

WIND

Before we can design to reduce wind speeds, we need to understand not just the prevailing (or main) wind direction but also any localised wind direction effects caused by landscape features, such as valleys and coastlines. At Abbey Home Farm, strong damaging winds tend to come from the southwest. However, extremely cold winds from the northeast in winter and spring can have a significant negative effect on vegetable crops as well. Placing tree rows across the wind direction will reduce wind speed. Conversely, it is actually possible to increase wind speeds if alleys are parallel to prevailing wind direction because they can act as a funnel, feeding the wind into the alley. It is not essential for tree rows to be perpendicular to incoming winds, however. A north-to-south direction that suits optimum light distribution to vegetable crops would also offer protection from both south westerly prevailing winds and north easterlies. We should add that if reducing wind speeds is a key objective, establishing good windbreak tree plantings around field boundaries is also an important integrated technique. You can also integrate windbreak trees at the end of alleys to prevent any funnelling effect of the wind.

Field Shape

Though we might be able to alter the shape of the field, the existing configuration is obviously a useful starting point for our design. When running a linear bed–style vegetable production system, there is an obvious advantage to orientating the bed to maximise the space available – usually beds running parallel to the longest boundary of the field. If positioning your tree rows along this same axis allows you to fulfil your objectives, then great. If not, then it may be necessary to rethink the layout of the field. This rethink may result in some parts of the field falling outside the alleys. Unlike arable farmers, who will tend to want to use the entire field for their one crop, this should not be a problem for vegetable growers. It is possible to make good use of small areas of field for crops that don't need a linear bed system, for instance perennials including rhubarb, soft fruit and asparagus. Other options would be using protected cropping and propagation structures or growing fertility crops for transportation to cropping areas, or putting the composting operation there. Last, but not least, the small field remainders can become permanent complex habitats for biodiversity, such as small woodlands, ponds and rough grass or wildflower areas. We don't see this as losing cropping area, rather as improving crop quality and yield through enhanced predator populations.

Existing Cropping Direction

If establishing agroforestry into a pre-existing vegetable system, there will be a current cropping orientation, often defined by the shape of the field. Even if the vegetable system is mature, do consider reorientating cropping to accommodate trees if this will make the system work in the long run. It is, though, a huge undertaking for well-established vegetable setups and is made even more complicated if no other land is available to move production to while reorientating. In this situation you may need to introduce trees in a staged manner over a number of years.

Topography

Topography is the physical shape of the landscape, including hills, valleys and rivers. At a singular field level, this can be viewed as slopes and undulations of the field. The shape of the land and the steepness of slopes are other fixed characteristics that impact on your design. Your topography may even be the source of some of the challenges you hope to alleviate with agroforestry. Soil erosion and water management, for instance, can be a problem in areas with adverse topography. Steep slopes can lead to significant topsoil erosion from rain, while the ground at the base or in valley bottoms can often be very prone to

Figure 7.7. Steep slope and erosion.

flooding. Conversely, steep slopes can be extremely free draining and exposed to drying winds in droughts, causing erosion from soil blow and adverse growing conditions for vegetable crops.

Topography has a strong influence on soil formation and can cause anomalies within fields or on flat ground, resulting in wet corners or patches or in changes in soil depth and type. If you know your site well, you should already have a good understanding of these quirks, but if you are taking on a new site, try to find out as much as you can before finalising any tree-planting decisions. One of our fields at Eastbrook Farm appears flat and uniform but actually drops 30cm (1 ft) or so from one end to the other, as well as shifting from almost pure clay soil at the one end to a chalky mix at the higher end. As a result, trees such as cherries that will survive at one end of the field have mostly died at the wetter end where the water rises almost to surface level in wet conditions.

The Keyline Technique

The keyline technique is a method of managing water at a landscape level that is applied in agricultural settings to manage water and soil fertility. It uses the natural shape of the land to define cultural practices, such as crop orientation, or cultivation techniques and directions. Initially developed by miners in the Harz region of Germany to manage water effectively for hydroelectric power, it was adapted and applied to agriculture by an Australian mining engineer-turned-farmer, P. A. Yeomans, in the 1950s. Championed mainly by organic farmers for many years, it is increasingly used across all scales of food production. Agroforestry has become synonymous with modern applications of the keyline technique, since the many benefits trees offer for slowing down water fit perfectly into the keyline technique.[1] In short, keyline design captures water at the highest possible point and transfers it outward toward the ridges using gravity, reversing the natural concentration of water in valleys.

Cultivation directions are deliberately contoured to the landscape to achieve this and, when this is complemented by parallel tree rows, the movement of water down the slope is slowed significantly. In this scenario, it is common for the tree rows not to be oriented north to south. Vegetable production might then be situated on south-facing slopes in the northern hemisphere, to capture sunlight. If you are

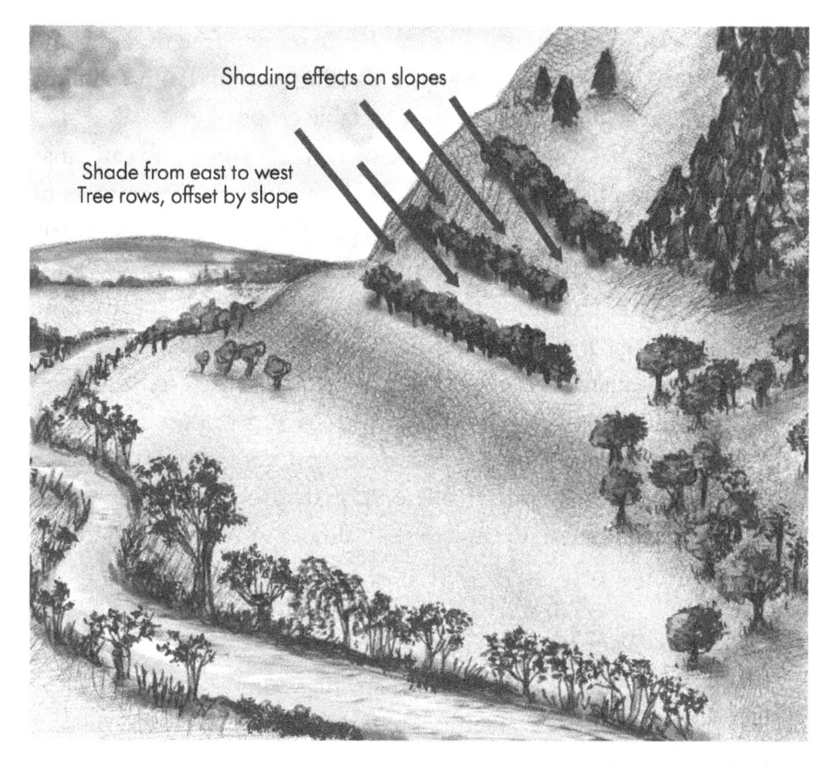

Figure 7.8. Shade on slope.

applying keyline technique to your slope, you might end up with tree rows running east to west. In this case, even though the negative aspects of shade on crops will be somewhat offset by the slope, it may still be worth applying the design techniques mentioned previously in the 'Aspect' section.

Balance of Aspect and Topography

It is very normal to feel confused and conflicted about design choices when weighing up the relative demands of aspect and topography. The best way to deal with this problem is to be led by your primary objectives. Let them define the aspect of your tree rows and then adjust the design to mitigate any negative results of the balance, such as excessive shading to the north of tree rows. Our advice generally is to prioritise the management of erosion on steep slopes rather than being too dogmatic over the points of the compass.

Understorey

Sustainable food producers dedicate time and effort trying to mimic natural processes, with agroforestry and the benefits of three-dimensional cropping being one great example of this effort. However, there is power, too, in the ground surface under the trees. We are talking here about the land left uncultivated under the tree canopy, so the tree roots are undisturbed. There is potential here to greatly enhance your system and help fulfil existing or introduce new objectives.

In a woodland, there are five distinct layers:

1. Canopy, consisting of mature trees with a range of heights
2. Low tree layer, of low-growing and immature trees
3. Shrub layer, with a mix of shrub species
4. Field layer, comprising grasses, ferns, flowers, brambles and other plants
5. Ground layer, with mosses and liverworts, along with seedlings of species found in taller layers

It is these distinct layers that are the key inspiration for forest gardens. Even in the simpler schemes that most growers will be designing, the understorey of a tree row is the perfect place to explore those principles a little further than is generally possible in the productive alleys.

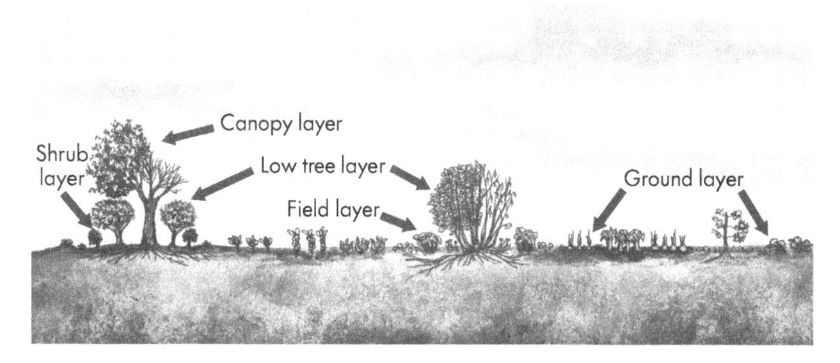

Figure 7.9. Canopy layers.

Understorey Options

Though our choices are more limited, there is still plenty of opportunity for introducing edible crops to the understorey. Though, in reality, there is probably a correlation between scale and complexity in this part of the design process. On a small scale, and particularly in gardens and allotments, there is an imperative to maximise production in every square metre but potentially also the time to manage a more involved design. On a larger scale, the understorey may still be productive but need to be simpler to allow efficient management. The key point is that each of these layers is exploiting a different ecological niche, and often will not overly compete with the other layers. In fact, they often complement each other and boost the overall productivity and health of the system. Depending on the complexity of our scheme, we can look to exploit all the layers or just select one to focus on.

Canopy

This layer will usually be defined by our tree choice. Edible examples of this layer would be walnuts (*Juglans* spp), chestnuts (*Castanea* spp), wild cherries (*Prunus avium*), as well as any fruit crops, such as apples (*Malus* spp), pears (*Pyrus* spp) and plums (*Prunus* spp), grown on original or vigorous rootstocks.

Low Tree Layer

Due to the dominant nature of true canopy-height trees, low trees are rarely used in vegetable agroforestry, except perhaps on very wide alley

Figure 7.10. Understorey options.

spacings or at the ends of rows. Therefore, the low tree layer often acts as a shorter canopy layer in an alley cropping system. For that reason many of the potential edible species in this layer will probably already have been considered in the design process. Examples include apples (*Malus* spp), pears (*Pyrus* spp) and plums (*Prunus* spp) grown on semi-vigorous rootstocks, as well as almond (*Prunus amygdalus*), peach (*Prunus persica*), nectarine (*Prunus persica*), medlar (*Mespilus germanica*) and mulberries (*Morus* spp).

Shrub

There are plenty of edible plants that can exploit this layer for growers of any scale. Examples are hazels (*Corylus* spp), elder (*Sambucus nigra*) and soft-fruit bushes, including blackcurrant (*Ribes nigrum*), redcurrant (*R. rubrum*), or gooseberries in general (*Ribes* spp), jostaberry (*Ribes × nidigrolaria*), or honeyberry (*Lonicera caerulea*). For the more involved intensive system, examples include trellised vining crops, such as kiwis (*Actinidia* spp) and grapes (*Vitis* spp), tayberry (*Rubus fruticosus × R. idaeus*) or loganberry (*Rubus × loganobaccus*).

Field

For the vegetable producer, this is the layer to explore perennial vegetables. Rhubarb, asparagus, wild garlic, Jerusalem and globe artichokes, as well as perennial herbs, will all exploit this niche.

Ground

This layer is trickier for larger-scale producers and is most at risk from competition with weeds, but it is definitely worth considering for home growers or smaller intensive systems. It is perfectly suited to short-supply-chain, high-value models. Examples at this layer include mushrooms, strawberries and edible flowers, such as nasturtiums (*Nasturtium* spp) and violets (*Viola* spp).

Flowers

Over the last decade at Abbey Home Farm, our eyes have been opened to the importance of cut-flower production. It is easy to consider food the most important product from the land, but we have learnt that we have an important role to play in feeding the soul. A bunch of flowers to a lifelong gardener who is fading away in a hospice or an agoraphobic trying to wrestle with their condition is almost as powerful and important as food for a hungry person.

There are many options for growing floristry products in your understorey. Most bulbs, such as daffodils (*Narcissus* spp), tulips (*Tulipa* spp), *Ranunculus* spp, and corms like anemones, will thrive beneath trees and capitalise on the light available to the understorey early in spring when the trees are yet to come into leaf. The same applies to coppiced stems, such as pussy willow (*Salix caprea*) and twisted hazel (*Corylus avellana* 'Contorta'). Once the leaves on the trees are out fully, it is shade-loving plants that thrive: *Alchemilla* spp, *Mertensia* spp, *Hydrangea* spp, toad lily (*Tricyrtis* spp) and *Aquilegia* spp, for example, while some are even more suited to deep shade, like ferns, Solomon's seal (*Polygonatum* spp), raspberry leaf (*Rubus idaeus*), *Hosta* spp, *Acacia* spp and holly (*Ilex* spp).

Planting to Encourage Predators

At Abbey Home Farm, encouraging predators was one of our key objectives as a tool for managing pests in our organic vegetable system, as well as on the fruit trees we were planting. Studies have shown that wildflower understoreys provide benefits to crops in alleys but also to the apple trees within the flowering strips. They are particularly effective at helping to control aphids, which are a significant challenge for both vegetables and fruit trees.[2]

As discussed in Chapter 5, the key to encouraging predators is to provide food and 12 months of undisturbed habitat. In terms of understorey, perhaps the best way of achieving this is to establish a perennial wildflower mix under your trees. This is what we did at Abbey Home Farm. We used a mix of more than 40 different species of annuals and perennials. You should select wildflower mixes that are suited to your soil type and system; for example, if you are undersowing a high-density tree planting, choose a shade-tolerant mix. Also remember that wildflowers left to stand through the winter play a crucial role in supplying wildlife with shelter and food. Birds, mammals and insects alike feast on the seeds dropped on the ground and seek overwinter shelter in the layers of dead foliage and dry hollow stems. Conventional wildflower management tells you to remove all foliage twice a year in the UK to maintain floral dominance and prevent grasses taking over. Our experience at Abbey Home Farm suggests otherwise. We sowed our mix three months before planting trees into it; we are now in year seven and not seeing grasses dominating, though we have witnessed a change over time as self-seeding annuals have become outcompeted by increasingly well-established perennial species. This kind of understorey will evolve over time as succession occurs and trees cast more shade over the understorey, which we discuss in more detail later.

Unmanaged Understorey

Perhaps the most powerful thing you can do to support biodiversity on any piece of land is absolutely nothing. We know that nature is perfectly able to look after itself when humans don't get involved. Therefore 'doing nothing' can be viewed as a proactive understorey option, not just an excuse for not having got round to it. Allowing natural succession to occur in the understorey can help maximise biodiversity if that is a key objective.

However, consider what access might be required for management of the trees, as understorey is likely to become fairly thick and impenetrable with species like bramble (*Rubus* spp) and dog rose (*Rosa canina*). If access is just required for early pruning and then harvesting after a long time or on long cycles, it would be manageable. However, if you are going to need to access the trees multiple times a year, such as when pruning and harvesting fruit trees, a wild, unmanaged understorey would become very inefficient.

Bare Ground

As with conventional vegetable production, some growers of tree crops have adopted understorey management methods using weedkillers, cultivation and mulching to eliminate all competition. Although weed control in some form is necessary in the immediate vicinity of trees for establishment when young (see 'Early Weed Management', page 137), it is not necessary to keep the whole tree row weed free at any point. In fact, this would be a massive missed opportunity to achieve many extra benefits.

Closely Managed Green Understorey

In many modern fruit production systems, the understorey is maintained as managed grasses and other low-growing plants, such as clovers and herbs. These are then cut mechanically with a mower, or grazed in more traditional orchard settings. Although managed green understorey is a common understorey in agroforestry, it is not often used with vegetables. Why not? Mowing is time consuming, and time is a valuable commodity for most vegetable growers. In an alley cropping situation, grazing could be an option, but fencing would have to be adhered closely to protect vegetable crops, though we can graze the understorey of tree rows when the alleys are in a fertility-building phase. There is also the risk of encroachment of grass into the cropped area.

If you do choose to go for a simple mowed grass understorey, then it can be managed with strimmers and pedestrian mowers. Or at larger scale there are a variety of mechanical mowers that can be purchased for tractors, quad bikes and two-wheel tractors. These are often designed for commercial orchards but are perfectly suited for many alley cropping scenarios. There are even mowers on a spring that will mow between the trees as well as the row, often referred to as swing mowers. The same equipment is being used for managing grass in solar farms, so this has led to rapid development and dropping in price of this style of machinery.

Preparation, Establishment and Management

Many times have we heard the phrase: 'I wish I had thought about the understorey before I planted the trees.' For those of you who have not planted trees yet – heed these words. Preparing for an understorey is best done in advance. Trying to prepare ground for an understorey after trees are planted presents problems. Preparing the whole strip (or even the whole field in new systems) means we can prepare the ground for both trees and understorey in one go. We can sort out any soil compaction or address fertility issues, but probably the most important is weed control. In particular, perennial weeds are perhaps the biggest challenge when growing any perennial vegetables without chemical weed killer. Once your trees and perennial crops are established there is almost no good way of removing perennial weeds.

A good long-term plan should involve a whole growing season of preparation for the tree rows, as this will greatly enhance establishment of both tree crop and understorey. Although we do not usually advocate spraying, there is an argument on sites that have very bad perennial weeds. The damage done from a one-off herbicide spray might be offset by the time, effort and plastic-use needed for a non-chemical approach. However, in most cases 12 months of good ground cover under mulches (for instance plastic, or paper/cardboard with wood-chip on top) or a growing season of bare fallow (where ground is tilled every four weeks over the growing season) will eliminate perennial weeds. Time spent on preparing your tree row will never be regretted.

The Long Game

There is a natural and inevitable evolution or succession in the understorey that you need to factor into any long-term design process. As trees mature, they cast more shade and so will affect the understorey differently over time. The edible understorey should be designed with this in mind, with you either choosing shade-tolerant species to start with or moving to them as needed. Wildflower mixes will naturally adjust as the shade increases. The roots of the trees will grow and fill the available soil in the tree row, dominating the competition for space, water and nutrients. This can be allowed for through tree spacing, density and thinning if required. Highly productive understorey crops will probably do best in the early years, when the trees are still small and uncompetitive. But if these are crucial to your operation, consider wider spacing of trees to allow room within the row for these crops to thrive too.

Root Management

In grass-based animal–agroforestry systems, it might not matter too much where the tree roots are growing, but for cropped situations, an understanding of how tree roots can negatively impact your vegetables is vital. Despite the general principles and assumptions around the productivity of diverse, multi-layered combinations, there are plenty of chances to get it wrong at the crop–tree interface. Some of the pioneering agroforesters in the UK have experienced these negative impacts as their systems mature. Wakelyns agroforestry, for example, saw their potato yields suffer from common hazel (*Corylus avellana*) rows planted too closely, while Tolhurst Organic have reported very competitive rooting behaviour from cherry trees (*Prunus avium*) impacting vegetable crops. At Abbey Home Farm, when we planted fan-trained peaches in our glasshouse, we soon had to root prune to prevent them from outcompeting nearby crops.

Not all systems will need pruning. You can leave wider tree strips to reduce the size of that competitive zone. You might also choose small enough trees that they will not have aggressive roots. However, with combinations likely to cause problems, it's worth thinking about how to manage roots right from the start. To achieve mutual benefit between tree and vegetable crops, you want to force the tree roots down below the vegetable root zone. This is done by severing the roots of the tree at the edge of the tree row to a depth below that of your vegetable crop roots. We recommend starting in year two or three after planting, and continuing the procedure on a regular basis, the frequency of which depends on the tree species and climate. In warmer parts of the world, agroforestry practitioners may prune every year.[3]

In temperate climates, root pruning may not be needed every year. Currently, there is little evidence to guide temperate systems on frequency of pruning; however, the risk of not doing so outweighs the cost and time spent on the operation. We would suggest at least every other year.

Table 7.2. Vegetable Rooting Depth Guide

Vegetable type	Maximum rooting depth
Lettuce	15cm (6 in)
Beets, broccoli, cabbage, carrot, peas, potatoes	45cm (18 in)
Parsnips, squash, sweetcorn	60cm (2 ft)

The next question is how deep to prune. In theory, this should be governed by the depth of the roots of the crops themselves. Most vegetable crops grown in temperate zones will have a maximum rooting depth of 15–60cm (6–24 in). Although, if they are actively fed and watered, they will not always reach the full depth stated in Table 7.2.

This is a lot shallower than most arable crops, which will regularly reach depths of 80–100cm (31–39 in). Assuming adequate nutrient and water supply to the crop, it would be advisable to prune to a depth of at least 50cm (20 in) for a vegetable system.

There is one further limiting factor to rooting depth, of course, and that is the depth of topsoil. When we grew at the Community Farm, some fields had nearly 2m (7 ft) of loamy topsoil, while at Abbey Home Farm we are grateful for the bits of the land that have 30cm (1 ft). This shallower depth will stop trees rooting out below the crops, but should also make it easier to root prune.

How you practically go about root pruning will depend on your scale, crop choice (both trees and vegetables), cultivation system and level of mechanisation. For a small scale 'no-dig' grower or market gardener using only hand tools or a walk-behind cultivator, we're afraid you'll need to get the spade out. If you have a lot of trees, then the laborious nature of digging a trench 50cm deep on a regular basis might make you consider using a physical root barrier; this is inserted in the trench before backfilling with soil. Options for physical barriers include metal, thick plastic and stone, anything that will remain impermeable to tree roots for a decent amount of time.

For larger-scale growers with tractors, it is easier. Any annual cultivations will by themselves break any tree roots present, although most sustainable growers, even if ploughing, tend to limit the depth of cultivations to 15–20cm (6–8 in). Therefore, it might be necessary to carry out specific root-pruning actions on top of your usual cultivations. Although some theorists talk about regular cultivating and weeding being enough, real-world situations with more rigorous species tend to indicate this is not enough.

In terms of implements, we need something capable of slicing or ripping roots at the required depth, which is run down the edge of the tree row. The most appropriate and easily available implement that will work at this depth is a subsoiler. Specialist root-pruning machinery

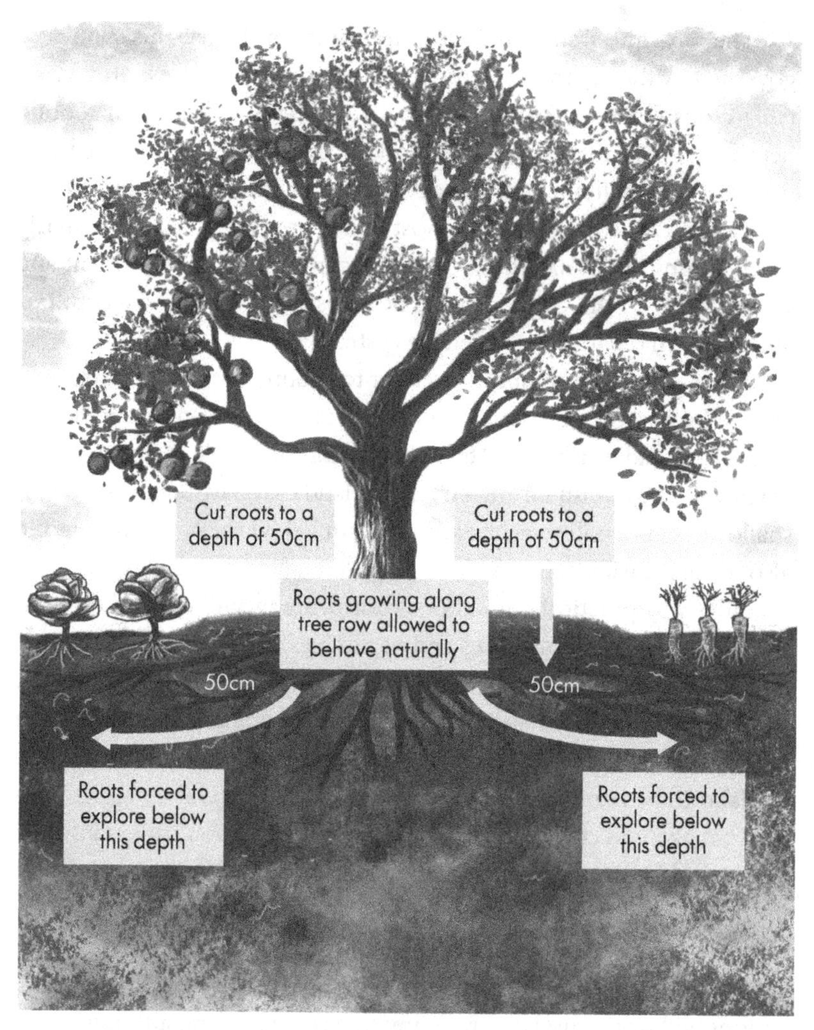

Figure 7.11. Root-pruning theory.

is widely used in the tree nursery industry and if available to use or rent is ideal. However, it is probably prohibitively expensive for most vegetable farmers to buy.

Rotations and Phasing

One of the things that often sets silvohorticulture apart from other forms of agroforestry is the need to fit in with complex rotations. If

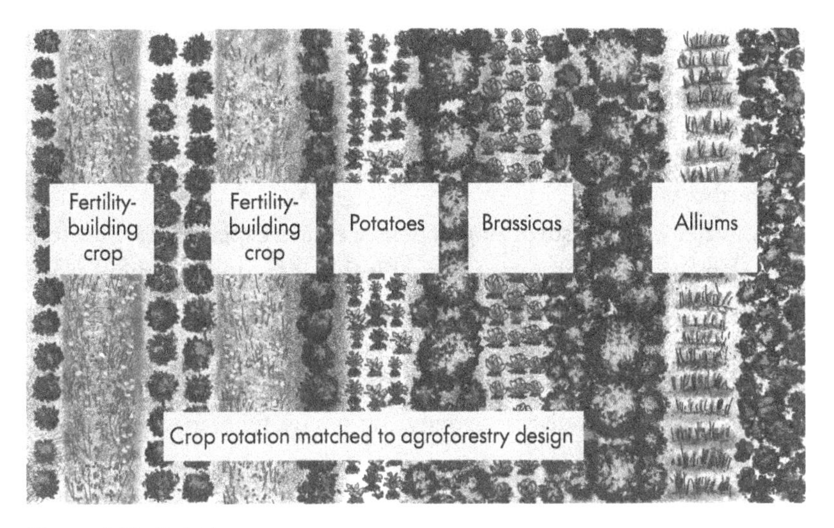

Figure 7.12. Rotation areas.

growers use coppiced trees, which are a common choice, this adds a further complication to the planning. We need to consider both seasonal and annual changes to tree height and density as well as the wide range of vegetables in our system.

Rotations

Crop rotation is an important and widely used technique in many forms of sustainable crop production. It is an invaluable tool for managing fertility, weeds, and pests and diseases in vegetable cropping. It consists of grouping crops according to plant families and fertility needs. These groups are then moved or rotated every growing season, with non-cropping fertility phases built into the rotation. Heavy-feeding crop families, such as potatoes and brassicas, often follow fertility-building phases and are in turn followed by less hungry crops such as onions, parsnips and carrots. We recommend matching your alley sizes to the size of your vegetable rotation area or making rotation areas divisible by alley size.

Having a physical barrier between rotation areas or within them has further benefits. Many pests of vegetables are low-flying insects, such as carrot fly, or ground-dwelling gastropods, such as slugs. In the case of the carrot fly, physical barriers are an effective tool to prevent their movement and a tree row with a well-developed understorey will provide a tall physical barrier. Although never 100 per cent effective, it is another tool in an integrated approach. As for slugs, the need to cross tree rows to reach other cropping areas will involve running the gauntlet of predators that live in the tree row, such as beetles and birds.

Another perhaps surprising benefit we have experienced with tree rows demarcating crop rotation areas is that it focuses our mind as we do crop walks. It seems that the compartmentalisation of crops with physical barriers has a positive effect on the way the grower looks at their crops, channelling attention to specific areas. With vegetable production being such a diverse activity, this is a useful management tool.

Phasing

Phasing applies to managing and exploiting different spatial opportunities over time. This is an activity associated with all the different disciplines of agroforestry and, in fact, in some scenarios it is a mechanism for a piece of land to evolve through the different disciplines of agroforestry over a long time scale. However, with vegetable agroforestry systems, phasing is an opportunity to take your design to the next level. It will add extra complexity to your design but in return will maximise use of time and space.

Phasing in the Tree Row

As trees within the tree row mature, they will become larger, both above and below ground. If your objectives are to provide a windbreak or reduce soil erosion, then you may wish to achieve this aim as quickly as possible; this can be done with close tree spacings and high-density plantings. However, as the trees grow it will be possible to fulfil these objectives with fewer trees. This allows the designer to 'catch crop' smaller trees or perennial vegetables early on in the system's evolution, before bigger trees shade them out. This would particularly apply if part of your tree row is for timber crops that prefer the competition early on to encourage straight trunk growth.

Phasing tree harvest is another opportunity to create growing conditions for certain crops, particularly suited to coppiced and pollarded systems, though it also applies to timber crops. When a tree is felled for timber or for woodchip, it immediately changes the growing

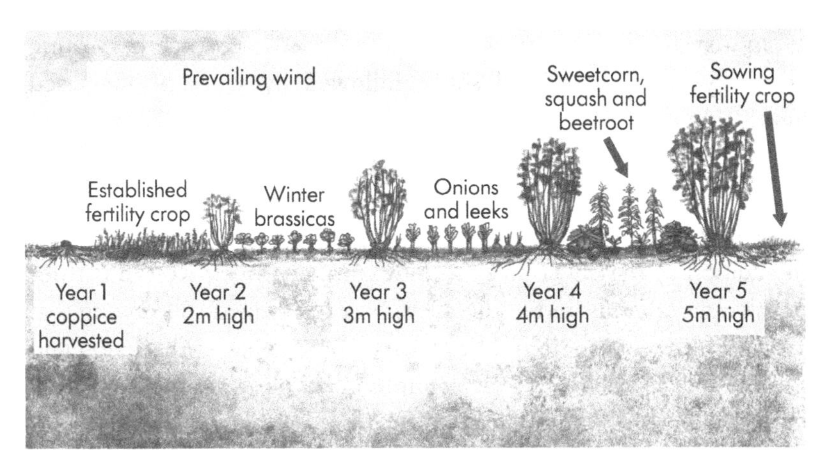

Figure 7.13. Abbey Home phasing.

conditions around it. For example, removing wind protection or allowing in more light.

At Abbey Home Farm, managing wind was a key objective for our vegetables, as was producing woodchip compost from coppiced trees. The way we have allowed for this in the design process is to cut the coppice in synchronisation with our vegetable rotation. Common hazel (*Corylus avellana*) and common alder (*Alnus glutinosa*) are coppiced on a five-year cycle. We have interplanted with trees for biodiversity, which will never be cut; therefore, there will always be some wind protection. Since we lose our maximum windbreak effect when we coppice, we harvest the row upwind of the area in the established fertility-building phase, which is the least wind-vulnerable part of the rotation. This allows the coppiced trees to regrow in time to provide full protection from the wind for vulnerable crops in the rotation. Squash and sweetcorn are particularly at risk from strong winds in cropping areas. Similarly, our maincrop, beetroot, is direct-sown in the field in June, when days are long and the weather often dry, not the best conditions for slow-germinating crops such as beetroot. The windbreak is an important part of increasing soil moisture and aiding germination. The same applies for establishing the fertility crop, again sown when soils have warmed up and days have lengthened. Fast and competitive establishment of fertility crops is crucial to outcompete weeds.

In a system with narrow alleys, the cycle of coppice harvesting creates years of increased shade followed by years of increased light. This can be synchronised with shade-tolerant and sun-loving crops accordingly.

Phasing in the Alley

In addition to the effect of phased tree management, we can further tweak our rotations within the alleys for each individual growing season. With the wide variety of crops used in vegetable production and the diversity of habit and seasonal timings of these crops, it is possible to capitalise on positive, and avoid negative, interactions between trees and vegetables.

A perfect example of this approach was adopted by Alan Schofield. Alan was an organic grower in the northwest of England and one of a group that helped to develop the first organic standards in the UK. He was a massive inspiration to us as both a vegetable grower and early pioneer of agroforestry in the UK. He ran his market garden based around an alley cropping system of pollarded poplars (*Populus* spp). He had a very intricate crop plan that made use of the levels of light close to the tree rows in an intelligent way. His brassica crops were laid out in the alleys according to how they coped with shade. Long-season brassicas, such as Brussels sprout and purple sprouting broccoli, were grown in the centre of alleys, as they do not like the shade, while shorter-season brassicas, such as kohlrabi and spring cabbage, would thrive on the edge of alleys despite the shade. In fact, he planted overwintered spring greens on the edge of alleys to make use of the lack of shade during the winter months and early spring. The year after the poplars were pollarded, he would grow squash and potatoes in that alley where there was increased available light.

Another way to view phasing in the alleys is over much longer time frames. For instance, if choosing narrower alley widths, as time progresses the light reaching the vegetable crops will slowly diminish as the trees get taller and canopies get wider. During the early years of the system, it would be possible to grow the full range of vegetable crops. However, as shade increases, we would modify the crop mix to include more shade-tolerant varieties. Then when the trees are mature, we might only grow shade-tolerant perennials, such as soft fruits and rhubarb. Perhaps at some point, the trees would be felled and the process started all over again.

In the southeast of the UK, some orchards would use the early years of establishing an orchard to crop vegetables between young trees to supplement income until the trees were cropping, at which point the vegetable production would stop and the focus would then be solely on the orchard production.

Phasing of the Whole System

It is only natural that we see the trees we are planting as a permanent new feature. Apple trees can easily live for well over 50 years, but their full commercial life is more like 20 to 25 years; after this fruit production can become erratic from year to year. With this in mind, we can even imagine the evolution of our agroforestry system over a very long time scale, mimicking the slash and burn techniques we outlined in the introduction. Whether you view trees as permanent or temporary features of your landscape depends on species choice and mix. Both approaches leave many phasing options to build into your plan.

On the one hand, you could envisage your land being in a long-term cycle of planting, establishing, cropping fruit trees and then removing and starting again, with different parts of the land being in one of these stages at any one time. On the other hand, a more traditional agroforestry system might evolve over time with the increasing size of trees. While the trees are young, the alleys are used for vegetable or arable production, and as the trees mature the system progresses into silvopasture.

CASE STUDY
Hilary Chester-Master and Andy Dibben, The Organic Farm Shop at Abbey Home Farm

Abbey Home Farm is a 1,600-acre mixed organic farm near the ancient town of Cirencester in Gloucestershire and has been proudly certified organic for more than 30 years. The current custodians of the land, Will and Hilary Chester-Master, share equal passions in nurturing biodiversity and feeding the local community with high-quality organic food produced free from chemicals and in sympathy with the land. The organic farm shop strives to supply the local community with a 12-month supply of 100 per cent organic local food produced at scale.

The farm produces 80 different vegetables, 15 different fruits, eggs, milk, beef, pork, lamb and a variety of arable crops including heritage grains, oats, wheat, barley and spelt. All produce is available in the farm shop throughout the year and available to eat on site in the lovely cafe.

TREE OBJECTIVES

With Will's enthusiasm for trees and Hilary's commitment to developing their range of local food, agroforestry had been on the cards for many years. Andy took over as head grower in 2015, and shortly after his arrival he worked with Hilary to plan and design an agroforestry system for one of the vegetable production fields.

The farm is a very windy site and reducing the wind was a key objective in our vegetable production areas. Trees are perhaps the most effective windbreak available. As organic farmers, we cannot and do not want to use artificial pesticides to manage problematic insects on our crops. We instead focus on nurturing natural predator populations to keep pest species at acceptable levels. We planned to use trees and their understorey to enhance our existing natural pest management (NPM) techniques.

With trees offering so many benefits to our vegetables, looking at what additional crops we could grow was strangely a secondary objective. Another core philosophy of organic farming is minimising the inputs that are brought on farm to produce crops. We were still buying a lot of propagation compost, so we saw an opportunity to provide the raw ingredients for propagation compost from trees. Although peat free for a number of years, we hope to now be able to go one step further and produce all our own compost from home-grown woodchip. We planted 660 common alder (*Alnus glutinosa*), white willow (*Salix alba*) and common hazel (*Corylus avellana*) trees in our system with this objective.

On the retail side of the business, we were buying in about 100 kg (220 lb) of apples a week to sell to our customers, so having decided to plant trees, growing apples was an obvious choice that we knew we could sell. We planted 100 apple trees of 13 different varieties, targeted at being self-sufficient each year between August and the end of February.

SYSTEM

With all of these objectives in mind we established an agroforestry system into a 5-acre field of vegetable production. We have two different alley cropping systems in use in the same field.

System 1 consists of just top fruit, mostly 12 different species of dessert apples on MM106 rootstock, with a few plums and cooking apples as well. The tree rows are 30m (98 ft) long and 4.5m (15 ft) wide with a single row of fruit trees down the middle planted with 5m (16 ft) spacing. They have an unmanaged grass understorey. There are 8 tree

Figure 7.14. Abbey Home Farm system.

rows spaced 18m (59 ft) apart. In the alleys, we crop eight 1.5m (5 ft) beds, with a bed left free for irrigation and wildflowers down the middle of them. Each alley is a rotation area. Crops include potatoes, leeks, pumpkins, cabbages and fertility-building crops.

System 2 consists of a mix of later-harvested varieties of dessert apples for storage, coppice species for woodchip production, and trees chosen specifically to enhance predator populations.

These tree rows are 100m (109 yds) long and 4.5m (15 ft) wide with a single row of trees down the middle planted on a 3m (10 ft) spacing, except the apples, which have 5m either side of them. They also have an unmanaged wildflower understorey. The alleys in this part of the system vary in width. Most are 50m (55 yds) apart as this matched our objectives for windbreak and predator range. We have some alleys of 36m (39 yds) and 18m (20 yds) widths. These are all divisible by the width of our irrigation boom, factoring an extra metre at the edge of alleys to allow boom clearance from young tree canopies. In the alleys, we run a five-year rotation of vegetable cropping, then two years of fertility building, followed by brassicas, alliums and then sweetcorn, squash and beetroot. We plan to harvest our coppice species on a five-year rotation, synchronised with our vegetable cropping rotation.

GLASSHOUSE

We have recently built a 1,000m^2 (1,196 yds^2) glasshouse on the farm and have established fan-trained peaches between some of the rotation areas. This serves two purposes for us. Firstly, we are prone to late frosts in the spring so would not risk growing peaches outside. Secondly, we are starting to regularly experience temperatures in the high 30s°C. This light intensity and heat stresses the plants, so the dappled shade cast by our peach trees will help to manage this.

Planting of all these systems took place over a number of years and allowed us to spread cost and labour requirements. This also allowed us to learn and modify designs each year with newfound knowledge.

CHALLENGES AND LESSONS LEARNT

Damage from wildlife has been a challenge through the establishment phase. Voles and mice were a problem when the trees were very small, and then birds landing on young branches and snapping them. After a few years, we removed the cage guards from the apple trees, but realised this was premature when deer damaged some trees, so we had to protect trees with modified tree tubes for another few years.

An early problem was creeping thistle migrating out from tree rows. We now manage this by weeding closely on either side of tree rows early in the season.

In the glasshouse, we were caught off guard by the growth rate of the peach trees, especially their roots, which had spread 3m (10 ft) in two years. This accelerated our planned root-pruning interventions, and now we plan to do this yearly.

THE FUTURE

We would like to properly establish our existing agroforestry systems before starting new plantings; however, as our knowledge and understanding of agroforestry increases and climate change intensifies, we see many scenarios where agroforestry might be a useful tool.

We would also like to explore further the possibility of growing perennial fruit in our glasshouse; apricots (*Prunus armeniaca*), grapes (*Vitis* spp) and kiwis (*Actinidia* spp) are all definite possibilities for the future.

A mixed row of apple and timber with a multispecies understorey benefits biodiversity at Abbey Home Farm. *Photo courtesy of Andy Dibben.*

Fan-training peaches to provide shade in glasshouse. *Photo courtesy of Andy Dibben.*

Fan-trained cherries with broad beans grown under plastic at Troed y Rhiw. *Photo by Nathan Richards.*

Fan-trained cherries with salad crops grown under plastic at Troed y Rhiw. *Photo by Nathan Richards.*

A coppiced hazel row (and a very free-range chicken) at Wakelyns Farm. They cut one half of the row each year. *Photo courtesy of Ben Raskin.*

Coppiced willow with cut branches waiting to be processed at Wakelyns Farm. *Photo courtesy of Ben Raskin.*

An almond and raspberry interplant at Eastbrook Farm shows the importance of weed control in the understorey. *Photo courtesy of Ben Raskin.*

Alan Schofield in his 5-year-rotation alder coppice agroforestry system. *Photo courtesy of Ben Raskin.*

Electric fencing at Henfaes Farm aims to confine chickens to the tree row for understorey management. *Photo courtesy of Ben Raskin.*

A high-density windbreak planting at FarmED, Gloucestershire, with deer guards and sheet membrane mulching. *Photo courtesy of Ben Raskin.*

Irrigating trees, understorey and vegetables at the same time using the systems approach at Abbey Home Farm. *Photo courtesy of Andy Dibben.*

Single row of trees mulched with woodchip at FarmED in Gloucestershire. A temporary electric fencing system protects against deer and sheep. *Photo courtesy of Ben Raskin.*

Cracks in heavy clay soil in a dry spring. The mulch is showing its worth, as the tree is still alive despite being right in the crack. *Photo courtesy of Ben Raskin.*

Ben cultivating close to newly planted apples in trained stockless system at Churchill Agroforestry Project. *Photo by Birgit Muller, Particle Productions.*

Tree seedlings heeled into woodchip pile waiting to be planted out. *Photo courtesy of Ben Raskin.*

A standard single apple row with managed grass understorey at Close Farm. On the left, fleece covers one of the mixed vegetable beds. *Photo courtesy of Ben Raskin.*

A mixed row of apples and hardwoods with forty-two species of wildflowers undersown at Abbey Home Farm. *Photo courtesy of Andy Dibben.*

An experimental apple and globe artichoke agroforestry planting at Riverford Farm. *Photo courtesy of Ben Raskin.*

CHAPTER EIGHT

Tree Planting and Establishment

We have now finalised and refined our design, including plans for the long-term evolution of the system. In essence we now move from the theoretical stage of the process to the practical aspects of planning the actual tree planting and establishment. However good our overall design, if the delivery is not well executed then it will not fulfil its full potential.

Preparation

A good crop plan is vital when you are in the middle of a hectic growing season and, as any vegetable producer knows, all growing seasons get hectic. This is no different with tree planting. When standing in a muddy field in horizontal freezing rain with hundreds of young saplings to plant and your team waiting for instructions on what to plant where, it is essential to have a good plan to follow.

Before a tree even goes near the ground, you need the site prepared, which might entail reorientating existing cropping direction. You will also need to prepare the ground itself well in advance. Traditionally tree planting happens in the winter. The reality of temperate climates often means that the window for cultivation is defined by the seasons. Cultivating wet soils is one of the most damaging things for soil structure, so mechanical ground preparation should happen in the preceding summer or autumn, definitely before the winter rains arrive. In the UK, this would normally mean before the end of September. The key purposes of ground preparation should be to control weeds, alleviate any existing soil structure issues and make planting quicker and easier. Depending on the situation, this might entail cultivating the whole field or just the intended tree rows. Remember this will be your last opportunity to efficiently and effectively carry out mechanical weeding of the tree rows. You might even mark the rows with a tractor and implement or create the slot (see the 'Planting Techniques' section on page 133), but again these need to be done while you can still get on to the soil without causing damage.

Planning starts by making sure all your resources are on site and ready to use. This planning includes trees being delivered on time and, if they arrive before planting day, having somewhere ready to heel them in. 'Heeling in' means you have somewhere to temporarily plant the early trees where their roots can be protected from drying and cold winds, and where they are safe from predation. Usually you will put them in a soil trench or in a pile of moist compost or woodchip. Get any mulching materials and tree protection on site and ready to position on the day of planting.

Planning labour can be tricky. For gardens and smaller quantities, you will be able to manage on your own, but for larger plantings you will need to find some extra help. Professionals can be expensive and not always easy to find, but they might still be the most cost-effective option because you know the trees are being well-planted and the professionals should be quick. Many of us rely on volunteers for planting, which can seem like free labour, but this requires much more supervision, including post-planting inspection. Volunteer labor also often results in the wrong trees planted in the wrong place, as it is almost impossible to keep track of who has what tree when you have a lot of volunteers. You also need to build in flexibility to account for weather, such as frozen or flooded ground, that might make planting impossible.

Timing

As mentioned above, the traditional tree-planting season is winter because the trees are not in leaf and are to a certain extent dormant. This time period allows them to adjust to their environment during dormancy and the soil to settle around their roots before spring triggers them into fresh growth. Most broad-leaved tree roots continue to grow during winter months, provided the soil temperature does not fall below 2°C (36°F).[1] Generally, the earlier we can get our trees in the ground, the better the odds of them surviving, particularly in drier soils. Late autumn planting is best for most species, as this gives trees a chance to become established before spring droughts. The Royal Horticultural Society (RHS) recommends early spring planting for wetter areas, as there is a risk of rotting in waterlogged soils over winter. Our experience is that trees planted by mid-February nearly always survive, while those squeezed in at the start of March will suffer if it is a dry spring.

Try to realistically assess how long it will take to plant your trees. The scale of the project, planting technique and available workforce will often define this. Be realistic and remember to allow time for fitting all planted trees with tree guards at the end of each day. If you haven't done this before, time yourself planting 10 trees, factoring in moving materials or getting more trees, as well as the actual planting. It is better to plant well than to plant fast. It may be necessary to plant over multiple days, so plan work accordingly. In fact, many good designs plan to carry out planting a system over several years because this helps spread the workload and cost, especially on larger projects. Spreading the planting can also add resilience. With the challenging and unpredictable weather patterns all farmers and growers now experience, it is impossible to guarantee good weather for the year of planting. A severe and early spring drought just after tree planting can lead to a lot of tree losses, especially if irrigation is not an option. Therefore, by spreading planting over several seasons, it is possible to spread the risk. It will also give you trees of varying ages, which is better from an ecological point of view.

Planting Techniques

Planting technique will often be determined by the size and type of tree you have sourced, as discussed in Chapter 7. Here are the most common techniques we use.

Notch Planting

Notch planting with a spade is the quickest method for planting small trees. It is generally suitable for small transplants. Sink a spade into the ground and move it backward and forward to open the soil. Then push the roots into this gap and push the notch shut around them with a firm heel of the boot.

In a variation of this, you cut a cross into the soil, then push your spade under the cross. As you lift the soil the cross opens up nicely, allowing you to pop your whip in and then stamp the soil back down.

Pit Planting

Pit planting is a slower but very effective way of planting trees and usually suffers fewer losses than notch planting. Suitable for larger bare-rooted transplants and whips, it is also the best method for planting root-balled trees and container-raised trees. Dig a pit just bigger than the roots of the tree. Then place the tree into the hole and backfill with the same soil. Gently move the tree up and down while returning the soil to the hole to ensure all air gaps are filled. Firm the soil in with the heel of your boot.

Other Techniques

Although the two techniques described previously are the most common planting techniques, there are many other ways of planting a tree.

On very wet sites there are various 'mound' planting techniques that aim to place the tree slightly above ground in a mound of soil to aid drainage around the roots. These methods include:

Turf planting: often used when planting small trees into very wet ground because it helps with drainage around the roots. Remove a square of turf and turn it upside down, then cut a notch through the turf and into the soil below. Insert the roots into the slot and then firm in.

Pit and mound: same as turf planting but used where grass is not present. The tree is planted into a mound of soil dug out of a pit next to it. Again, this greatly assists drainage.

Ridge and furrow: the mechanised version of the two previous techniques. Use a plough to turn over a single furrow and plant the trees into the inverted strip of soil. The furrow allows for better drainage around roots.

If using any of the three mound techniques, be sure to mulch quickly. If the weather is dry, having trees planted on raised mounds leaves them very exposed to drying, which is why these techniques are best limited to very wet sites.

Figure 8.1. Notch planting.

Figure 8.2. Pit planting.

Mechanisation

On larger plantings it is naturally tempting to speed up the process by using machinery to help. This often involves using a subsoiler or transplanter pulled behind a tractor to open a slot in the soil and then placing trees into the slot either by hand or from a trailed planter. Although this can be an effective technique, be aware that in dry weather the slot can open back up after planting and allow the roots to be exposed to the air, leading to poor establishment or loss of trees. At Eastbrook Farm, we have used a subsoiler in this way and it speeds up planting significantly and in theory helps to break the soil structure a little, easing root establishment. However, on our heavy clay, we had one year with significant losses. This can be guarded against by well-timed and adequate organic mulching, and by planting a few centimetres to the side of the slot so that if the ground splits open the transplant is not right in the gap.

Planting Depth

Trees should be planted at the right depth. You should be able to tell by the mud on the stem what depth they were positioned when grown at the nursery. Aim for the trees to be sat at this same depth once soil has settled after planting. Some species, like willows (*Salix* spp) and poplars (*Populus* spp), will be very forgiving, while others, like conifers, are a bit more fussy. For grafted fruit trees, it is absolutely critical that soil does not get mounded up at the base of the tree and bury the rootstock. This will cause the upper part of the grafted tree, 'the scion', to start rooting itself, which will bypass any effect of the rootstock.

Root Pruning

Many species of tree can be quite heavily root pruned at planting. This won't be necessary for small transplants and whips but can be a useful technique for larger bare-rooted fruit trees. We have used this method extensively for apples and pears, though we would be wary of using it on *Prunus* species as there is greater disease risk. We learnt this approach from apple guru Matthew Wilson, who trims the roots down almost to a single stem. The two main benefits are that one, you have a much smaller root mass and therefore need to dig a smaller hole. And two, his

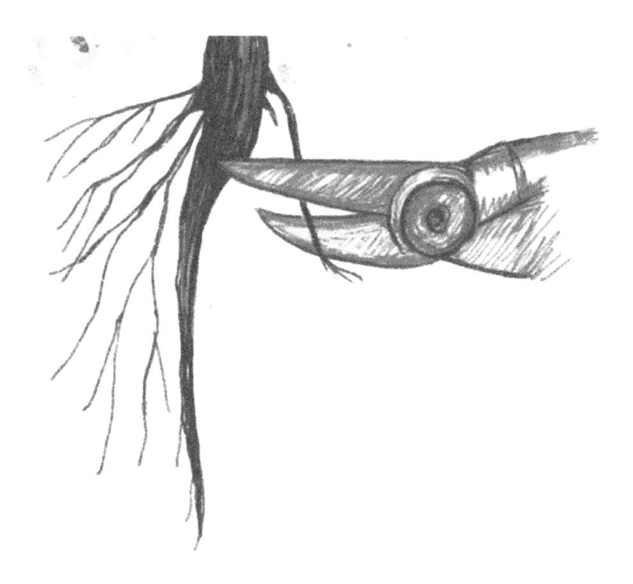

Figure 8.3. Root trimming.

theory is that the cuts stimulate root growth immediately out from the stem rather than risk the circling of roots that can happen in pit planting.

Additives at Planting

There are plenty of 'planting aids' you can buy, from fertilisers to mycorrhizal treatments. We have had experience with some of these, and broadly we would suggest that, if your soil is fertile and in good health, they will add little to the fortune of your trees. However, if planting into degraded or heavily cultivated soils, or in shallow and lighter sandy soils, adding biochar, compost or an organic fertiliser can significantly improve success rates. If you have access to leaf mould from a nearby woodland, then adding a little of this to every tree hole might be the most effective way of inoculating with the largest range of locally adapted microorganism species.

Early Weed Management

As a vegetable grower, you'll know all about weeds. They have the potential to outcompete crops for water, light and nutrients, leading to low yields or crop failure. These basic principles are the same for trees. Address perennial weeds well before planting if you can. Not all are problematic. For instance, we have one field at Eastbrook Farm where creeping thistle (*Cirsium arvense*) is rife in some rows. It dies right down in winter, creating a mulch, so after the trees were established in year three we haven't worried too much about trying to remove them, though clearly seeds from these plants can be a problem. Annual weeds can mostly be controlled by mulching or by establishing a good understorey, as discussed in Chapter 7 (see the 'Understorey' section on page 111). Controlling weeds in the early life of the plant is vital for good establishment, so don't allow problem weeds to go to seed. If, despite your best intentions, weeds do become a problem, we recommend a tight mowing regime for a season to regain control.

Before Planting

As mentioned earlier in the chapter, the only time you can effectively weed the tree row mechanically is before the trees are planted. This is not an opportunity to be missed and is without a doubt the best time to tackle any pernicious perennial weeds such as docks (*Rumex* spp), creeping thistle (*Cirsium arvense*) and couch grass (*Elymus repens*). Bringing the roots and rhizomes of these weeds up to the surface and desiccating them in the sun and wind is the best non-chemical approach. This method takes time and the right weather, and ideally is allowed for in the summer preceding planting. The advantage of addressing weeds in advance of planting is that it allows you to deal with both weeds that will affect your trees and also those affecting your understorey. On a small scale, it may be possible to use some of the weeding techniques outlined in the following to manage weeds in the understorey, as well as round the bottom of trees. However, on a large scale this becomes very expensive and resource hungry.

At Planting

At the time of planting, concentrate on the area directly around young trees to help them establish and receive adequate light, water and nutrients. The aim should be to effectively suppress weeds around the tree to 50cm (20 in) from the base for the first two to three years. There are a number of options for achieving this.

ORGANIC MULCH

Organic mulching of trees serves many purposes: suppressing weed growth, conserving soil moisture, modifying extremes of soil temperature, improving drainage, supplying nutrients to the tree, feeding soil life and improving soil structure. The technique is simple and involves heaping organic material around the base of the tree. It is important that the mulch is thick enough to suppress weeds but does not pile up around the base of the tree, which can create damp conditions and bark rot. You will want to either top up the mulch for the first three years or apply a thick enough layer at planting that it will suppress weeds for at least two to three years. The literature suggests at least 10cm (4 in) depth for effective mulching. At Eastbrook Farm, we have discovered that 25cm (10 in) is not only more effective but also reduces the need for subsequent mulching. Since mulching is an expensive and time-consuming business, if we can get away with only doing it once we will.

There are many choices of organic mulching materials, including woodchip, cut grass, straw, hay, sheep wool and compost. Which option you choose will often be defined by what is readily available, cheap and easy to apply. Fresh woodchip will give the longest-lasting weed control.

The easiest mulch to apply is organic materials that have been made into matting, which comes in rolls or as individual mats. They can be made of hessian, cardboard, felt, paper, hemp or jute. These manufactured materials can be effective but are also expensive and in our experience can encourage voles by providing an attractive environment around the tree's roots. For more details on mulching, see *The Woodchip Handbook*.[2]

SYNTHETIC MULCH

Synthetic mulches include plastic sheeting, landscape fabric and rubber matting. This option is often attractive because they are relatively cheap, easily available, and quick and easy to use, particularly on larger-scale projects. However, the effects of microplastics on the environment are well documented and should be considered. If choosing to use synthetic mulches, one should use and manage them with the environment in mind. It is particularly important to plan and carry out their removal and recycling at end of life or use. There is generally a choice between permeable woven fabrics and impermeable plastic sheeting. The woven fabrics tend to be tougher and more expensive, and they allow water and air to reach the soil, which is better for soil life and tree health. However, they are considerably more expensive than plastic sheeting.

Impermeable plastic sheeting is cheap and easy to use but is very vulnerable to damage by wind and wildlife. Using narrow strips will allow more water and air to reach the roots of the tree and soil.

Whichever option you use, ensure they are applied effectively at planting with adequate anchoring along edges. This can be done with staples, stones or soil. Then, over the first few years, regularly inspect the mulch: check anchoring is still secure, remove any weeds growing through or across the mulch and any organic matter build up on top of the sheeting. When the sheeting either starts to deteriorate or is no longer needed or effective, immediately remove and if required replace. Most trees will need weed control for only the first two to three years.

Mechanical Weeding

Mechanical weed control around trees is regularly used in orchards and nurseries. However, be aware that trees often have shallow roots near their base that are important for gathering water and nutrients, so mechanically weed at a sensitive depth and distance from the tree. This depth and distance will change as the trees mature; in early years it may be possible to effectively manage weeds close to trees mechanically with shallow cultivations. However, to effectively manage all weeds, especially in-row weeds and weeds near base of trees, specialist and expensive equipment is often required for mechanical weeding.

Herbicide

As organic growers, we don't use herbicides ourselves but recognise that they are a very cost-effective way of controlling weeds around tree planting. However, there will be some damage to soil health, and, more importantly, you get no moisture retention or temperature modulation benefit as you do with mulch.

David Granatstein, an American orchard researcher, completed a comprehensive study of different weed control methods in commercial orchards that use compost and woodchip mulches. Although the most expensive to implement, these mulches more than paid back the outlay cost with quicker tree establishment and higher overall fruit yields.[3]

Tree Support: Stakes and Ties

Smaller trees may not need staking, though the guards we use to protect them might. However, most fruit and nut trees are likely to need support for the first years of their life. Our usual advice on stakes is that whichever size stake you were thinking of using, get a bigger one. A lot of stakes you buy are not great quality and break or rot before the tree is ready to cope on its own. We suggest a stake of at least 6cm (2 in) in diameter. If you can source a hardwood stake, such as sweet chestnut (*Castanea sativa*) or black locust (*Robinia pseudoacacia*), it will last much longer. The stake should ideally be between one and two thirds the height of the tree, though you can of course prune to reduce tree height and use a smaller stake. A shorter stake allows the tree some sway, which helps it to strengthen its trunk. However, trees that have been 'forced' in high-nutrient nursery settings can be a little weak when planted in high-wind situations, so for high-value trees we now tend to opt for a slightly taller stake.

There is a myriad of options when it comes to ties. Unfortunately, most of them are plastic. We are experimenting with hessian ties, which tend to rot after a year or two, but that is probably better than not getting round to loosening a plastic or even rubber tie, which can strangle a tree. Despite best intentions, it is amazing how quickly the trunk thickens and, even within a season, trees can suffer. If you can, stake on the windward side of the tree; this helps prevent the tree being blown into the stake, which can cause rubbing.

Some systems require a permanent trellis or wired support system. Vining species – such as grapes (*Vitis* spp), kiwis (*Actinidia* spp) and loganberry (*Rubus* × *loganobaccus*) – will all need long-term support. However, using wires to support trained forms of apples (*Malus* spp), gooseberries (*Ribes* spp) or stone fruits is also a great option for smaller-scale growers. These systems are highly productive and the wire trellis allows for narrow fruit rows at easily pickable height. There is a higher establishment cost, but this may be paid back by increased yields from a very small area.

Irrigation

All trees need water to establish. Whether this is supplied naturally by rain or from supplementary supplies will vary hugely from site to site and from season to season. Geographical location and climate, soil type, topography, access to water, tree species and rootstock choice will all play a role in defining whether and how much you irrigate your trees. However, droughts are predicted to become more frequent and intense as our climate changes, so if you are relying on rain it is worth considering and planning for potential droughts. As vegetable growers, mostly we have the ability and infrastructure to be able to water plants, but even if you intend to irrigate it is wise to try to minimise water requirements, and to ensure that you have enough to supply both trees and vegetables in that critical spring period.

Spread the Risk

As discussed earlier in the chapter, spreading planting over several years can offset the risks posed by droughts to young trees. It is also sensible to have a variety of species in your design as this, too, will spread the risk of extreme weather impacts.

Avoid Late Planting

Timing of planting is important for drought resilience. Ideally, new plantings get a good soaking by winter rains before spring, which ensures soil settles tightly around roots and reduces air gaps, preventing the risk of roots drying out and dying. It also allows roots to establish quickly, start growing and begin to be able to source water from a wider and deeper area of the soil. Later plantings risk not receiving enough water to properly establish, leaving them vulnerable to spring droughts.

Conserve Moisture

As described under weed control, reducing competition is important for conserving water. Mulching greatly reduces evaporation of water (typically by about 25 per cent)[4] from the soil and is the most effective technique for reducing drought risk to young trees. Mechanical weeding can increase evaporation of soil moisture, so time this activity well to avoid any risk.

Cultivations can be used to assist with water conserving. In drought conditions, soil surrounding trees can become baked solid, leading to water running off the surface when it does rain. So gently cultivating soil in alleys between trees to assist water infiltration is a technique used in some dry areas.

Tree Choice

Species choice is important for establishment but also long-term water requirements. Thirsty trees can struggle to establish in droughts, and many high-value fruit crops, whether tree or bush, will require good levels of moisture to supply high yield and quality. Rootstock choice is also important. Dwarfing rootstocks produce fewer fine roots than the more vigorous rootstocks, making dwarfing rootstock more vulnerable to lack of moisture.[5]

Irrigation Options and Techniques

Knowing when and how to water young trees will help with fast and healthy establishment and, if water is limited, will ensure it is used in a targeted and effective manner.

Critical Time Periods

Providing adequate water over the first two to three years of planting is vital for good establishment, but there are also key times within one season for successful establishment and production. Fruit-bearing trees produce fine water roots throughout the spring and summer, and require water throughout the growing season, whereas non-fruit-bearing trees tend to produce most of their fine roots in May and June, meaning this is a more important time to provide adequate water. As with vegetable crops, targeted timing of water at flowering greatly prolongs the potential pollination period, leading to higher yields in fruit crops from trees.

Technique

When establishing any crop, it is better to supply larger amounts of water infrequently. This encourages roots to grow and colonise deeper into the soil. If we irrigate little and often, trees will tend to root shallower, making them more reliant on irrigation and vulnerable to drought.

As discussed throughout the book, adopting a systems approach is important and this particularly applies to irrigation. Often vegetable growers will have irrigation systems that can be adapted to irrigate trees. Using existing equipment and allowing design to accommodate this will lead to more effective use of water and time. At Abbey Home Farm our system was designed to allow tree crops to be watered at same time as vegetable crops.

Options for irrigating tend to be either drip systems or sprinklers. Sprinkler systems are more effective for establishing trees but require higher volumes of water and higher water pressure to deliver. Drip irrigation can be effective to help trees through droughts when establishing and also to supply targeted water at key times as discussed previously. On a smaller scale, a good soaking with a hose (or even watering cans) at appropriate times is effective, but don't plan to do this on a large scale as it is prohibitively time consuming.

Tree Protection

Trees have successfully managed to grow for millions of years without humans protecting them; however, humans have destroyed the balanced ecosystems that allowed that to happen. A direct result of this has been the persecution of top predators, especially in farming landscapes. The lack of predators – such as wolves, birds of prey, wild cats, otters and foxes – has led to an uncontrolled rise in population of their prey species. Deer, rabbits, voles and squirrels can all have significant impacts on food crops. Young trees, too, can have their shoots, roots and bark eaten, leading to poor establishment rates and often failure. Some damage may be tolerable, but this will depend on your objectives. In most situations, we need to provide some protection, but deciding how much is mostly a balance between cost and risk.

Whether to Provide Protection

A good initial assessment of the site will flag potential pests. The presence of large numbers of deer, rabbits, hares or voles indicates that protection will be needed. You will probably know if these species are present on your land, but if you are not sure then visit the site at different times of day and see if you can spot them. Dawn, dusk and night are their most active times. Look for damage to existing trees and vegetation. What kind? What height? Can you see footprints or droppings to help identify them? Trail cameras are a useful tool, too.

Once you have identified they are present, see if you can work out where and how they are getting in and out of the field. Walk the boundary and look for obvious entry and exit points.

Assessing the presence of voles can be harder but surveying rough grassland areas in the field can help. You need to look for burrow entrances into rough grass, runways under the grass, droppings and sometimes plant clippings where they have been feeding. It is also worth checking with local environmental groups, farmers and residents who may have good local knowledge of wildlife.

Deer

Red, sika, fallow, roe and muntjac deer are present across the UK, with different regional variations in population for each species. With perhaps the exception of some inner-city locations, deer will be a potential challenge for most tree-planting projects. Deer damage young trees by browsing, fraying and bark stripping. Larger deer are more damaging because of their higher reach and greater feeding demand. The size of the population will define the amount of damage but, if left unprotected, a small number of deer can cause a lot of damage to young trees very quickly. Fencing and tree guards are the standard way of managing deer. The height of protection needs to be matched to the size of the deer species, as does the gauge of fencing (see details in Table 8.1 on page 147).

Rabbits and Hares

Populations of rabbits and hares vary. Due to disease, they are less numerous than they were 20 to 30 years ago; however, they can still be a significant problem for tree establishment. They will eat buds, shoots, growing tips and bark off young trees and can also expose roots when burrowing. It is likely that, in vegetable systems, you are already controlling or excluding these animals, but if not, shorter guards will keep them off your trees.

Voles

Voles are common inhabitants of many fields and gardens, often found in rougher, unmanaged areas. They are, in fact, a key indicator of a healthy habitat to ecologists, and an incredibly valuable source of food for birds of prey. They are also a common problem for young trees, chewing and damaging bark at the base of the stem, as well as chewing roots. Trees may recover from minor damage but will die if ring barked, meaning bark is removed from around the whole circumference of the tree. Damage can occur at any time of year, but often peaks in late winter to early spring. The most effective protection is using a tree tube and ensuring the tube is pushed into the ground to a depth of 10cm (4 in). A woodchip mulch applied directly on the soil around the base seems to be an effective way of discouraging tunnelling around roots; however, sheet mulching, both biodegradable and synthetic, can encourage voles because it offers them good protection from predators, especially if laid on a whole tree row.

Protect the Field, Tree Row or Individual Trees

The answer to this question will be defined by the scale of the project, the available budget, the existing boundary and the kind of protection required. The positives and negatives vary with every site. An existing boundary fence or hedge can be restored and managed to meet requirements. For example, a well-laid thorn hedge is very capable of keeping out small deer but is of no use for rabbits. Rabbit netting added to an existing livestock fence will manage hares and rabbits (though they tend to find a way through in a couple of years), but will be of limited use against deer. Adding a whole new boundary fence is often prohibitively expensive, especially on larger projects, but grants may be available to help with protection costs.

An important factor to consider regarding boundary fencing is the impact on the wider movement of wildlife, as described in Chapter 5. Biodiversity is a powerful tool for food producers if allowed to flourish and should generally be welcomed into cropping areas and not excluded. Badgers, whose trails are used by successive generations, are well known to take exception to boundary fences and will happily tear holes in them to reopen their routes. This can be managed with badger gates, but the same philosophy should be applied to all animals. So, from a biodiversity perspective, protecting tree rows or individual trees is often preferable to protecting an entire field.

There are an increasing range of products on the market for protecting trees, but the basic principles for protecting from different animals still apply. Many farms will be able to recycle existing materials, so

Table 8.1. Guard and Fencing Heights for Different Pests

Animal	Fencing	Tree guards
Large deer – red, sika, fallow and roe	minimum 1.8m (6 ft) high with maximum 150mm (6 in) square mesh size	red, sika, fallow – 1.8m (6 ft) high roe – 1.2m (4 ft) high
Small deer – muntjac	minimum 1.5m (5 ft) high with maximum 75mm (3 in) square mesh size	1.2m (4 ft) high
Rabbits and hares	minimum 1m (3 ft) high with maximum 31mm (1 in) hexagonal galvanised mesh netting; bottom 300mm (1 ft) of height turned outwards away from the planting (towards the rabbits)	rabbits – minimum 600mm (2 ft) high hares – minimum 750mm (2.5 ft) high
Voles	It is not possible to manage voles with fencing.	200mm (8 in) high, pushed 10mm into soil

Table 8.1 provides basic parameters to be applied accordingly, whether recycling or purchasing.

Tree Guards

Broadly speaking tree guards come in two forms, tubes and cages.

Tubes are small-diameter sleeves, varying in height, and secured with a single wooden or metal stake. The stake should be positioned upwind of the tube, to prevent damage to trees in high winds.

The advantages of tubes are that they are:

- easy to weed and mulch around
- cheap
- able to create a helpful microclimate for establishing young saplings
- quick to remove at the end of life

The disadvantages of tubes are that they are:

- a brake on saplings becoming stable against wind
- less robust than cages

- able to provide an ideal habitat for voles and mice, leading to damage to young trees
- unsuited to very hot weather and can cause trees to suffer heat stress.
- often made of plastic, which presents end-of-life and environmental impact issues (Biodegradable options are available on the market but currently expensive.)

Cages vary in design and, therefore, costs vary widely. They offer opportunities for unique designs and material selection and are often adaptable to budget. Cages are commonly constructed of metal or plastic mesh and secured using wooden or metal stakes. Mesh size and gauge are the more important characteristics to consider. Heavy gauge with larger mesh size would be appropriate in areas with significant deer pressure. Conversely, smaller mesh size and lighter-weight material would suffice in areas with heavy rabbit pressure. Often protection against both is necessary, so wrapping the base of large cages with small-mesh material meets both requirements.

The advantages of cages are that they are:

- more robust and longer-lived than tubes
- able to allow saplings to become 'wind strong' sooner
- much better suited to recycled materials and easier to reuse multiple times
- less habitatable for voles and mice
- more suited to fruit trees than tubes because they allow easier access for early formative pruning

Disadvantages of cages are are that they are:

- often labour intensive to install and remove
- harder than tubes for weed control around base of cages
- often considerably more expensive than tubes if purchased new

Livestock

There are many advantages to integrating livestock with vegetable production: ducks eat slugs; chickens eat leatherjackets and can terminate green manures effectively; pigs can clear out old crops, such as potatoes and carrots, and will eat weeds as well; sheep help manage cover crops and resting land; and they all fertilise the land. Longer-term, livestock can be a useful tool for managing the tree understorey. However, when establishing your trees, you should probably try to keep animals away from them. Tree guards are an effective way of achieving this if they are robust enough and of an appropriate size. They must be able to prevent intended livestock from reaching through the guard to the tree.

Electric fencing of the tree row can be a quick, cheap and effective way of providing protection. We have used permanent, electric mains-fed single strand to keep cattle off the trees, but many other systems are available. The big advantage of movable or temporary electric fencing is that it can move with the livestock and not be in the way of your vegetable operations and machinery when you don't need it. The style of electric fencing must be appropriate for intended animals.

Fertility

Trees are autotrophs, which means that they feed themselves by making sugar in their leaves through photosynthesis. But they require adequate levels of nutrients to help fuel this process. In nature, trees get nutrients from organic matter, air, beneficial microbes such as fungi and bacteria, and soil minerals. Well-established trees should be able to do this without much help if situated in a healthy soil. When establishing young trees, the focus again should be on soil health and structure, as much as individual nutrients.

Before Planting

This is a key opportunity to address any soil structure issues and nutrient deficiencies. Where applicable, preparing a site with a cover crop before planting might be the most effective technique for improving soil structure and maintaining organic matter and nutrient levels. Effective and timely termination before tree planting and establishment of understorey is important.

At Planting

Even when situated in healthy soils, many young trees will benefit and establish more vigorously when provided extra help in the first couple of years. A good, healthy application of organic mulch as described previously is the most effective way to address both soil structure and fertility in the early years. If applied to the surface of the soil, it will fuel soil life, which will enhance soil structure. Also, the mulch breaks down slowly over time, releasing a wide variety of nutrients to the tree over the early years.

Although it can be tempting to add compost or nutrients to the planting hole, this should be avoided as it can result in tree roots not wanting to establish beyond the planting holes, as the presence of compost around their roots reduces the need to search further afield. The slower release of nutrients from a surface applied mulch ensures a balance between providing nutrients and encouraging natural root spread.

After Planting

Although established trees should be able to feed themselves, it is still important to maintain a healthy soil structure to allow them to do so. Compaction from machinery is a risk and needs to be avoided; 80 per cent of soil compaction happens with the first pass of the tractor. Compaction can also be a problem after a major flooding event, so in these situations an application of organic mulch can be a good way of alleviating the repercussions of compaction. Root pruning with a subsoiler will also help to break any surface compaction.

Once again, cover crops are an effective way of maintaining soil health and nutrient levels, and for this reason are a valid understorey option. Plus they will also be able to boost biodiversity, if managed appropriately. Combining low-growing legumes and deeper-rooting grasses is an effective way of maintaining both soil structure and soil nitrogen levels.

Understorey

The basic guidance for establishing the understorey is the same as establishing the trees themselves. Preparation of ground, timing and technique of planting, weed management, irrigation and fertility all need to be considered.

Wildflowers and Cover Crops

We recommend preparing the ground well in advance of planting, ideally the summer before because this will be your only opportunity to address perennial weeds. Weeding so early will also deplete some available nutrients, which can be advantageous for perennial wildflowers.

Sowing the understorey is best carried out before planting of trees, especially when using machinery. Early autumn is a good time, to allow for germination and early establishment of hardy annuals and perennials before winter. There tends to be less slug pressure than in spring, and it also ensures soil is covered for winter to maintain good soil structure and health.

When winter arrives and it is the time to plant trees, the action of planting trees and mulching will immediately destroy weed seedlings near the trees, and, when spring arrives, both trees and understorey are ready to flourish in warm rains. Establishment is everything with this kind of understorey and poor establishment can be hard to remedy after trees have gone in, often leading to weed problems.

Edible and Floral Crops

If perennial crops are grown in the understorey, then really focus on weed management, both before and throughout the life of the crop. Fertility is also an important consideration for perennial crops. Like trees, they will benefit from good levels of organic mulch to aid establishment. It is also possible to grow perennial crops, such as soft-fruit bushes, with green manures such as clovers to help with fertility. Unless you are using machinery to plant this understorey, planting can happen at the same time as or after the trees. If growing trellised crops, such as blackberries or vines, installation of support structures is far easier before tree planting.

CHAPTER NINE

Early-Years Management

Despite our best intentions, trees often die in their first few years. While there are some things you can't protect them against, sadly neglect is one of the leading causes of young tree death. Making time to inspect and attend to young trees diligently over the first five years is crucial. Once-a-year inspections are the absolute minimum, and in the first year we would suggest trying to look at all your trees more often, particularly when planting on a new site or if this is your first time.

Mulching

Trees like good, rich soil around their base. At maturity they get this through dropping their own leaves and through the action of soil organisms, which build soil organic matter. However, when they are young, and especially when growing out in the open and not in a nurturing woodland environment, this may require human intervention in the form of organic mulching.

As discussed in the previous chapter, mulching fulfils a number of positive roles for young trees: suppressing weeds, conserving moisture and feeding soil life. If you have applied a substantial amount at planting, often this will be enough. In some circumstances though you may need to top up the mulch.

Here are a few signs to indicate a supplemental mulch is needed.

Roots not well covered by soil. This sign is particularly important in the first couple of years after planting because soil will settle around the roots, and wash down into the cracks and air pockets caused by planting. This soil movement can lead to roots becoming exposed. If this has occurred, then a top-up of mulch will protect the tree roots.

Extreme weather events, such as flooding or drought. Extreme weather has led to cracking or washing away of the soil. This, again, can lead to tree roots becoming exposed, and a good helping of mulch will protect them.

Weed competition. Mulching is an effective and sustainable method of controlling weeds. However, it is only effective if maintained at a weed-suppressing depth. Even if applied at a good depth upon planting, over the following years the mulch becomes thinner. How fast this occurs depends on many variables. For example, an exposed site can lead to mulch being blown away when dry, and a sandy soil naturally low in organic matter will consume mulch faster than a rich soil. The type of organic mulch will also define how long it takes to break down. Green compost is unlikely to last even a year, and woodchip from willows (*Salix* spp) and other fast-growing soft-wood species will have mostly broken down after one season. But hard woodchips, such as oaks (*Quercus* spp) or hornbeams (*Carpinus* spp), can last up to three years. The size of the chip also has an impact on the speed of disintegration. As a guide, you should aim to keep an area of 1 square metre (11 ft^2) around the tree weed free for the first three years. However, if the tree is growing slowly or the weeds are so big that they appear to be competing with the tree, then we would recommend additional mulching beyond that period.

Signs of drought stress. The signs – typically small, discoloured or shrivelled leaves – can show either due to extreme weather or from being outcompeted by a vigorous understorey.

Signs of nutrient deficiencies. While most trees are tough enough to establish themselves without any added fertility, if leaf or soil inspection and analysis shows deficiencies, mulch can be part of the solution. At Eastbrook Farm in years two and three after planting our almonds, we noticed what looked like magnesium deficiency. Leaf analysis confirmed this. Research revealed that this deficiency is not uncommon in this species, particularly in the latter part of the growing season. We decided not to treat but just to top up with mulch and see what happened. We have not seen any problems since and are now in year seven.

We are very interested in exploring the properties of single-species woodchip mulches. For instance, some early work looking at willow (*Salix* spp) woodchip suggests that the salicylic acid in the bark can stimulate the immune systems of apple trees, to reduce the severity of the damaging disease scab (*Venturia inaequalis*). Researchers at a Soil Association Innovative Farmers' field lab, working with Dr Glynn Percival of the University of Reading, looked at how willow chip might work on apple trees in the field. Willow varieties differ in their levels of salicylic acid, so the field lab looked at different varieties of willow and how much mulch was needed to be effective.[1] Much more work is needed to fully understand how to harness this power.

If you use sheet mulching, whether synthetic or biodegradable, check that organic matter does not build up on top of the material, which allows weeds to grow. If there is a lot of organic matter and weed growth on top of synthetic mulches, one way to deal with it is to flip them over. Do also check underneath sheet mulch materials for evidence of voles and mice. You might see the actual creatures, but also look out for early signs of damage. As soon as the plastic mulches start to deteriorate, remove and responsibly dispose of them.

Tree Protection

Tree protection is probably the area that will require the most attention over the early years of a tree's life. If you can make regular checks throughout the year, and especially after strong winds, you can nip any problems in the bud before they cause major damage. Guards can work loose in the wind and then rub on trees, causing damage to trunk or branches. Loose guards are also less effective at protecting trees because small animals can sneak through them to nibble the bark, and larger ones might be able to push them aside for a browse. Plant material can build up within the tube or spiral guards, which can lead to moulds developing, or ants and other creatures moving in. A common problem with solid tree tubes is mice using them to build nests from balls of dried grass a foot up the tube and living there for the winter. During this time, they will slowly gnaw through the stem of the young tree. We have had experience of this in our systems, so check the tubes in early winter for nests, and ensure that tube guards remain firmly pushed into the ground, as this helps with protecting the stem from vole damage and also stops rabbits getting their noses underneath.

Remove the guard at the right time! Every tree grows at a different rate, and different guard designs have different tolerances and ability to grow with the tree, so it is not possible to say exactly which year to remove the tree guard. The most important factor is to remove the guard before it physically restricts the growth of the tree. Spiral guards if not removed in time can become absorbed into the tree, causing long-term damage. Tree tubes can restrict the growth of trees if not removed in a timely manner, although the latest designs have a perforation to allow the guard to split when the tree becomes too big. Even if they don't fully constrict the tree growth, they can become abrasive as they break, causing rubbing damage on stems and branches. As emphasised in the previous chapter, timely removal of plastic tree guards is not just important for the health of the trees but imperative for the health of the ecosystem.

One advantage of many cage guard designs is that they can be undone, meaning you can loosen or adjust the guard as the tree grows. If the cage is of solid design, ensure you have removed it before the canopy is too big, or you may have to come back with an angle grinder to get it off. When growing young fruit trees in cage guards, it is worth pruning with cage removal in mind. Branches left to grow out through the cage will make cage removal without damaging the tree a lot harder.

It is not uncommon for trees to outgrow their initial protection but still be vulnerable to pests. In this case, we'll need to replace the old with something suitable for the larger tree. At Abbey Home Farm, we had alder trees that quickly outgrew their tree tubes, but when removed the trunks kept getting eaten by deer, so we installed cage protectors.

Other factors may force a change in how we protect trees. For instance, at Abbey Home Farm we found that some of our young trees on a windy site in an open field were being blown about too much. Effectively the guard was acting as a sail. We are now coppicing some of our willows and alders in year six, much earlier than we planned, to get them to grow back multi-stemmed, which is more robust and less vulnerable to the gales. We will protect these with cages.

Another time we removed cage guards from fruit trees only to find the deer started to eat the trunks. We dealt with this by reusing old tree tubes that had split. We wrapped two around the trunk and secured them. Doing this for a couple of years enabled the trees to grow to a size where they could fend for themselves.

If you have fenced the whole field, don't forget to maintain the fencing. Rabbit fencing, particularly, needs a good inspection regularly. At Eastbrook Farm, we deer- and rabbit-fenced one of our fields. Feeling confident, we left the trees unguarded within the field. For a couple of years all was well, but in year three we started to see some bark damage in one area of the field. Checking under an area of brambles that had sprung up along one section of the fence, we found that rabbits had found a way in and were using the protection of the thorny patch to carry out tree munching raids. Brambles (*Rubus* spp), dog rose (*Rosa canina*) and grasses can all lead to rabbit fencing being pulled down or gaps appearing under the fence. Badgers are capable of tearing holes in rabbit wire, too, as described in the previous chapter. If deer fencing has been used, always inspect it after strong winds, especially if installed along woodland, to ensure branches or trees have not fallen on it, compromising your protection.

Overall, with most tree plantings, as with sustainable vegetable production, you are looking for the trees to reach the point where they don't need protection. There are some occasions though when this won't happen. When coppicing for timber products or woodchip, it is common to cut the tree down to almost ground level. This leaves the new growth exposed to browsing damage by deer. At this stage the deer damage does not usually kill the tree but it will affect the quality of the product being produced. Eating the growing tip results in stems branching, which is undesirable for those end products where straight timber is required. For example, hurdles, bean poles, building materials and cricket bats all require straight-grained wood. Traditionally, in woodlands coppice regrowth was protected from deer by woodsmen piling large amounts of brash (wood trimmings created when harvesting stems) on top of the tree; the regrowth would grow up through this pile as it rotted, and the deer would avoid it as they do not like sticking their heads into piles of brash. The other way to protect regrowth of coppice is to alter the cutting height to a point where your local deer cannot reach, as discussed in the 'Pollarding' section on page 160. It is also possible to use cage guards for a few years to allow tree regrowth to get beyond the reach of deer.

Deer can also cause problems on small mature fruit trees, eating young buds and growing tips, even if they can't get to the main trunk. This is best addressed through pruning, as discussed below.

Tree Support: Stakes and Ties

Inspect stakes and ties at least a couple of times a year, especially after storms. Stakes need to remain firmly in the ground to provide adequate support. If they work loose – in other words, if you can rock them back and forth at all – then they will either need to be knocked further into the ground or, if that is not possible, pulled out and resecured. Sometimes, the quickest option is to knock another (thicker) stake in next to them and tie the two stakes together. Many modern soft wood stakes don't last that long and, if they rot before the tree is big enough to cope on its own, you will need to replace them. There is the potential to spend a fair amount on restaking, which is why we recommend using really good stakes to start with. Although it is expensive it will likely save you time and money in the long run.

Tree ties are the easiest thing to miss and a leading cause of death in young trees after neglect. Even within a season, the girth of a tree can expand significantly, and trees that look like they had plenty of room to grow in the spring might be severely restricted by the autumn. Regularly adjust straps to accommodate growth. Rubber ones have more flexibility to expand but tend to be less strong. At the same time, check how stable the tree is without the support. Ideally, we want to remove the support as soon as we can to promote a stronger rooting system. However, if you remove a support and the tree starts to fall over, slowly during a long period or quickly after high winds, you may need to restake.

If growing fruit trees or vines on a wired system, that will need checking, too, as the ground can settle around the posts and extended dry weather can lead to uprights becoming loose. Wires and cables can also loosen over time and may require tensioning.

Pruning

As a vegetable grower, you may only have pruned and trained plants such as tomatoes, peppers and cucumbers. The subject of tree pruning might feel daunting, with so many different types of tree and so many end uses for any one tree, all potentially requiring different pruning techniques. This book is not the place to try to cover all of this information. Suffice to say the first step is to establish what we are aiming to achieve with our pruning. For many trees, especially those not intended for cropping but as tools for biodiversity, soil erosion or windbreaks will happily achieve our desired objectives with no pruning at all. As an example, pruning out and removing dead wood might be sensible in a productive fruit crop but will have a negative effect on biodiversity, as decaying branches and trunks play a critical role in ecosystems. A gnarled, twisted and deformed tree will often provide a crucial niche habitat for our rarer species, so only prune when the end product or outcome requires it. Broadly we would suggest there are three main reasons to prune:

- annual production of fruit, nuts, flowers or foliage, both formative and ongoing pruning
- coppicing or pollarding on a multi-year cycle for woody products and biodiversity
- forming a tree for timber production

Pruning for Annual Production

Whether producing fruit and nuts, or something more unusual like *Hamamelis* for flowers, we need to do some formative pruning in the first years to help produce the right shape for our production system. There is then ongoing pruning aimed at maximising production and maintaining the harvest numbers in a balanced, quality product from year to year. As trees mature, there might also be a need for some restorative pruning. When building an agroforestry system around old or neglected trees, this restorative approach is crucial to getting them back into good production.

Early pruning helps to form a good framework of mature wood. Exactly what is required depends on species, production system, age of the tree at planting and how much the nursery has already done. Some trees, especially those on vigorous rootstocks, will be delivered with long leaders. These should be pruned back, to both encourage new growth in the direction you want and prevent wind-induced rocking and leaning. We had some plums at Eastbrook Farm that were so healthy and vigorous that they had nearly two metres of growth at delivery. As we wanted a good-sized tree, we made the mistake of not pruning them hard enough. We had also staked them low, a technique that can work well to build strength on shorter trees. Sure enough, winds over the following year blew the overly tall stem constantly and we ended up with leaning trees. We spent the next three years trying to straighten them with extra stakes and harder pruning. Being braver with our pruning at planting would have given us a better tree with less work.

Once you have your shape, which might be a goblet, a pyramid or a trained system like espaliers, ongoing pruning aims to promote fruiting growth and allow for good levels of light and air movement around blossom and fruit. We recommend specific research on how best to prune individual species for maximum production. As a rule, winter pruning (for species that can be pruned at that time) will promote growth, while summer pruning will reduce vigour. Summer pruning is a great way to control the size of trees in intensive production systems. The challenge, of course, is that the labour needed comes at a time when other vegetable crops also need a lot of work. This is why we are interested in growing species like apples (*Malus* spp), pears (*Pyrus* spp) and quince (*Cydonia oblonga*) on a coppicing or pollarding cycle. Yield would be zero in year one after pruning, but if a proportion of trees are cut on a three- or four-year cycle you would always have some trees fruiting each year.

Coppicing and Pollarding

Many hardwood trees, if cut to the ground, will grow back with multiple stems. This ability of trees has been harnessed by humans since the stone age, allowing us to manipulate natural growth of woodlands to provide building and farming materials, firewood, gunpowder and more recently fuel for power stations. Stems are cut in winter on different yearly cycles according to tree species and desired end product.

COPPICING

To create coppice trees, we establish young trees to a point where their roots will vigorously fuel the regrowth of multiple stems when the tree is cut to the ground. For a very vigorous species, such as basket willow (*Salix viminalis*), this could be as early as the first winter. For other willows (*Salix* spp) and poplars (*Populus* spp), winter two or three is appropriate. For trees, such as alders (*Alnus* spp) and hazels (*Corylus* spp), a five-year cycle is recommended. The regrowth from this cut is then left until stems achieve desired diameter for your objectives or it is time to harvest in accordance with your phasing plan as discussed in Chapter 7. For example, willow catkins for floristry would need all stems to be harvested every spring. Whenever cutting coppice, always angle the cut to shed water toward the outside of the stump to prevent water collecting and causing rot in the stump centre. Cutting was traditionally carried out with a tool called a billhook, a curved heavy blade, but hand saws and chainsaws are often used. In large biomass operations, mechanical harvesters have been developed.

POLLARDING

Pollarding is the same process as coppicing, but, rather than cutting the tree at ground level, it is cut higher up, often above head height. It is a technique employed in landscapes where livestock are a feature, as it protects new growth from their browsing. However, this same technique can be employed effectively to prevent browsing by deer. In an agroforestry scenario, pollarding can offer the opportunity to increase space beneath the tree for understorey use and vegetable activities in the alley, and it is especially relevant in a mechanised system. It can also be a way of allowing more sunlight to reach the north side of trees if your rows are oriented east to west.

Timber Trees

The aim when pruning for timber is to create a distinct leader that will give the tree a strong, dominant leading shoot and so go on to produce a straight tree. A straight tree is required to produce stable timbers that dry evenly and have uniform strength along their lengths. It is common in forestry to use the competition of other trees to encourage straight growth and reduce side shooting. Growing trees for timber in an open environment can have advantages in that the trees grow more quickly, and your financial return is quicker. However, pruning becomes increasingly important, to maintain uniform straight growth and to prevent or remedy forking of the growing tip. It is also important to remove lateral branches and dead branches that can create knots and therefore weaknesses in the timber. The end destination of the timber produced will define the quality required and therefore the rigorousness of the pruning regime. The general aim is to maintain a clean trunk with a good canopy for photosynthesis at the top.

Weeding

Weed competition will slow the establishment and vigour in young trees, even if they survive it. As the tree grows, this risk lessens. If we have planted and mulched our trees well, the first year or two should be OK, but it is easy to then forget and allow the weeds to take over. Some weed species, such as creeping thistle (*Cirsium arvense*) or docks (*Rumex* spp), may look a mess but will not cause much harm to trees and will die back every year, providing organic matter to the trees. Other weeds, like brambles (*Rubus* spp) or couch grass (*Elymus repens*), are more dangerous and should be controlled. Try to keep the immediate vicinity of the tree weed free through hand weeding and mulching, though mechanical options at scale as previously discussed are also an option provided the small trees and fencing allow access.

Perhaps a more often faced problem in a vegetable-based system is migration of problem weeds, such as creeping thistle, dock and couch grass, from the tree row into vegetable cropping areas. In the case of docks and thistles, this can be through seed dispersed by wind. The best way to manage this, depending on scale, is early pulling of weeds or a well-timed cut to prevent seeding. Creeping thistle and couch are often more of a challenge in alley cropping because they are rhizomatous, meaning they can spread through replication of roots alone. They are specialists at colonising bare ground, so are perfectly suited to moving out into alleys from tree rows. This began to be a problem with our system at Abbey Home Farm, so for a few years we left a bed empty either side of the tree row and kept them weed free all season. Now we use these beds but plant them with late crops to allow for early season weed control prior to planting. This kind of weed pressure is often a challenge with an unmanaged understorey designed for biodiversity, so in this scenario managing weeds prior to planting is essential to prevent the spreading problem.

In no-dig systems, perennial weed migration from tree rows into cropping areas is a particular risk. Root-pruning trees on a small scale could serve the purpose of both managing tree roots and also preventing weed ingression into cropping areas. Deep mulch-bed systems and deep paths filled with fresh woodchip, currently very popular in market gardening, might also hinder weed encroachment.

As the trees mature and become the dominant plant in the area, they will outcompete and suppress weed growth immediately around them. But be aware that if trees are to be regularly harvested, such as coppice varieties, the sudden increase in light after harvest years will trigger seed germination and plant growth that may lead to an unforeseen return of weeds.

Why Integrate Trees and Crops?

On the very simplest level, trees are brilliant. Since they evolved 360 million years ago, they have been a central element in the evolution of all life on Earth, providing habitat and valuable resources. The removal of trees from farming landscapes has had subtle but profound impacts on both production of food and also the wider environment. It is time we reintegrated them into all our food-producing landscapes.

To the vegetable grower, trees offer a great opportunity for increasing the variety of food grown in the same field or garden, while bringing other benefits to your crops. Wind protection, soil health improvement, flooding offsets, drought management, soil erosion reduction and shade are the key benefits that can be exploited by introducing trees. These benefits might be tools to address inherent historical problems on your site, such as soil erosion, wind or waterlogging. They will also be relevant as climate change takes a grip, helping to solve problems you have not experienced before. Whether you have already been majorly affected by climate change or not, now is the time to future-proof your vegetable production from these challenges that are sure to affect all food producers over the next twenty years.

What Can Trees Do for Me?

We have made some bold claims about the potential benefits of trees. How this potential is realised is truly down to treating each site as a unique opportunity. Each field and garden demands a unique design to truly maximise benefits. Identifying current and potential problems on your own site is the single most powerful step of the design process. Take your time to understand and assess this. Discover other benefits, such as new crop lines and fewer farm inputs, on top of this initial assessment. Many trees can fulfil the problem-solving role of system design, and more often than not a diversity of tree species will be needed for this role.

The truly creative part of system design combines solving your specific site challenges with providing a usable end product. This product might be a new crop line that you can sell – edible, floral, medicinal or drinkable. Often overlooked, however, is the financial and sustainable impact of reducing your production inputs. Once you factor this in, a whole new level of benefits can be realised. Propagation compost, general purpose compost, predator population boost, pollinator population increase, trellising materials and irrigation reduction are all a result of growing trees, without which you would see us adding to the burden of what is already a resource-hungry activity.

Witness the environmental impact of your inputs by producing them locally. Reduce the financial impact of your inputs by producing them yourself.

Which Trees Fulfil These Objectives?

Having identified the problems you wish to solve, and the new crops and farm resources you can produce on site, you need to choose which species will fulfil one or many of these objectives. How you choose and lay them out in the field or garden is the truly creative and exciting part of the design process, translating theory into reality. The individual qualities of different tree species, such as rooting behaviours, canopy shapes, heights, rootstocks and growth rates, are all things to become intimately aware of, not just in one moment but as they evolve over time. When considering sellable crops, you have to understand if they will thrive here now or in the future, and if they can be sold. When trying to reduce farm inputs, you also need to work out if you can produce meaningful amounts of a particular resource.

How Will the Trees Interact with Crops and Time?

Having decided which trees to plant and how to arrange them, the fine tuning of the design process is understanding how this arboricultural intrusion is going to interact with your vegetable crops. To achieve this, you need a realistic grasp of the complex interactions between different tree and vegetable species. Balancing competition for light, water and nutrients is the key; and understanding how these interactions happen over a seasonal and multi-annual time scale is crucial. Extrapolate your design over a 10-year period and visualise every year, trying to run different scenarios of what real-world incursions are possible: predation, market changes or intensifying weather for instance.

When Does Each Stage Need to Happen?

Plan. Prepare. Procure. Plant. Protect. Problem-solve.

And most importantly, ENJOY.

Annual vegetables are an exhilarating yearly addiction, enslaving many vegetable growers.

Adding the long game of trees to your system results in a deeper and more reliable, resilient and profound presence to your production system.

Life is short. True happiness lies in enjoying the moment but keeping half an eye on the future.

A philosophy the authors have bonded over and both see embodied in agroforestry of all forms, a passion it has been a pleasure to share.

APPENDIX

Tree Directory

We cannot hope to cover all the possible tree species that might be used in a vegetable agroforestry system; however, we have identified some that we hope will help you make decisions about what to include in your own system. They are mostly from our own experience and so do have a UK focus. Many of these species are found around the world, and we do include some that are not native to the UK but that we feel have some potential to perform well, particularly with a changing climate.

We have also chosen five 'key species'. These are trees that, for us, perform such a range of functions within a cropping system that we would always consider them in our design. This does not of course mean that you have to; there may well be others that you think fill similar niches and would work better for you.

Pollinator rating: We based our pollinator rating mostly on how much pollen and nectar is produced. Wind-pollinated species will tend to rank lower, even though they provide valuable food as pollen and sometimes honeydew, because they do not provide nectar. For us, the flowering time is also crucial, particularly including species that flower either very late or early to help provide food for as long a season as possible.

Woodchip: While we can make woodchip from any tree species, and we might well chip the prunings or thinnings from a range of trees, we have listed woodchip as a product for only those species that we are likely to grow specifically if producing regular quantities of woodchip is our objective. These therefore tend to be quicker-growing coppiceable species.

Forage: This book is primarily for vegetable growers, but we are aware that there are mixed systems that include livestock within the rotation or in parts of the holding, so we have indicated which species might provide forage for animals. Even on a home-garden scale, these can be cut and fed to individual animals.

KEY SPECIES

Common name(s)	Typical height	Pollinator rating	Fruit/nut
Latin name	**Useful attributes**	**Flowering months**	**Other saleable output**
Hazels	12m (39 ft) fully mature coppiced: 3m×3m (10 × 10 ft) on a 5-year cycle	low	hazelnuts and cobnuts: both whole nuts (in and out of shell) and when processed into a range of added-value products larger wholesale market: round-nutted varieties
Corylus spp	coppiceable	February to March	bean poles: for use on farm or for sale firewood: a 10+ yearly coppicing cycle floristry: catkins or twisted hazel cultivar walking sticks: small market woodchip: denser and harder wood than willow, perfect for mulching supply of products into hurdle making, thatching and building
Willows	up to 25m (82 ft), depending on species coppiced: from 2m×2m (7 × 7 ft), according to species and length of rotation	high	none
Salix spp	coppiceable	basket willow (*S. viminalis*): February to March pussy willow (*S. caprea*): March to April crack willow (*S. fragilis*): May	firewood: chipped biomass floristry: catkins woodchip: quick growing and high yielding with potential to boost immune responses of mulched trees medicinal

Biodiversity benefits / Risks	How to fit into agroforestry system
Catkins are a very early source of pollen for insects. The pollen is available at night and so is important for night-flying moths. Hazelnuts also feed a variety of birds and mammals, like the red squirrel and bank vole. The dormouse is also closely associated with hazel and is a keen eater of caterpillars. squirrel control for nut production	Coppiced: Hazels grow approximately 50cm (20 inches) per year, so after 5 years it could be nearly 3m (10 ft) tall. After 10 years it will likely have reached 6m (20 ft), and by year 15 might have grown to 9m (30 ft). Choose a rotation to give the height of the tree that fits your crop spacings and final use. For instance, ramial woodchip for soil health would suit a short cycle, while bigger logs for wood fuel would need a longer one. For nuts: If nut production is your primary objective, then you will probably not coppice but instead grow rows of larger nut-producing trees at greater spacing. These could be integrated with smaller coppiced rows on either side to optimise the shelterbelt effect.
Supports a wide range of species with more insects living on willow than other tree species. Early pollen in spring for bees to build their colony. water competition invasive roots near drains or other plants	Coppiced: More suited to siting away from main growing areas and using the coppiced woodchip for boosting soil health and productivity in cropped area. Boggy areas: Good for helping to dry boggy patches by improved infiltration and water uptake.

KEY SPECIES

Common name(s)	Typical height	Pollinator rating	Fruit/nut
Latin name	**Useful attributes**	**Flowering months**	**Other saleable output**
Alders	10m (33 ft) as a mature tree coppiced: 4m×4m (13 × 13 ft) on a 5-year cycle	low	no
Alnus spp	coppiceable (However, some species, such as rubra, are less reliable. A higher stool or even pollarding is recommended to improve results and regrowth.) nitrogen-fixing	March	firewood rot-resistant timber: much of Venice built on alder piles and poles for riparian repair[1] woodchip
Apples	up to 6m (20 ft), depending on rootstock, soil and climate	high	fruit, with added value through processing
Malus spp	coppiceable	April to May	firewood: small quantities produced from pruning mature trees timber: specialist timber market for woodturning when felling mature trees
Cherries	3–8m (10–26 ft)	high	cherries, from domesticated species
Prunus spp	quick-growing timber species	April	firewood timber: high value from 30–60-year-old trees

Biodiversity benefits Risks	How to fit into agroforestry system
Alder catkins provide an early source of nectar and pollen for bees, and the seeds are eaten by siskin, redpoll and goldfinch. Foliage is food for many moth and butterfly species, including the alder moth.	Nurse crop: Quick growing so good for a nurse crop for slower trees. Woodchip: Use the chip produced to support further plantings or build soil health.
Phytophthora spp diseases: a threat in many parts of the world some species invasive, e.g. common alder (*A. glutinosa*) in US	
Provides an excellent source of nectar and pollen, especially if different varieties are grown to extend the flowering period. Useful for building up colonies of all bees in spring, with surplus honey being stored if the weather is good. Supports a range of insects and spiders. Excess fruit is a valuable food source for many species to feed before the winter.	Apples are a supremely versatile tree that can be grown at almost any size and trained to suit the vegetable-growing setup. For growers with their own outlets, the fruit is readily saleable so is usually financially useful, too.
scab and mildew (in damp climates)	
Spring flowers provide an early source of nectar and pollen for butterflies, bees, bumblebees and other insects. The fruit are eaten by birds, including the blackbird and song thrush, as well as mammals, such as badger, wood mouse, yellow-necked mouse and dormouse. The foliage is eaten by caterpillars of many moth species.	Interplant: Mix with other species to produce a shorter-term timber crop or use as main fruiting species. Biodiversity: For those with biodiversity and natural pest management as key objectives, this is a high performer.
suckering highly prized by birds – netting is crucial in most situations, which can be tricky and not cost effective in many agroforestry situations, so plan carefully	

FRUITING

Common name(s)	Typical height	Pollinator rating	Fruit/nut
Latin name	**Useful attributes**	**Flowering months**	**Other saleable output**
Pear	up to 10m (33 ft), depending on rootstock	high	fruit
Pyrus spp	coppiceable	April to May	firewood timber: small specialist market
Plum, greengage, damson	up to 4m (13 ft)	high	fruit
Prunus domestica	potentially coppiceable	March to April	firewood timber: small specialist market
Peach/nectarine, apricot	up to 7m (23 ft)	high	fruit
Prunus persica, P. armeniaca	potentially coppiceable	March to April	firewood timber: small specialist market

FRUITING AND NITROGEN-FIXING

Common name(s)	Typical height	Pollinator rating	Fruit/nut
Latin name	**Useful attributes**	**Flowering months**	**Other saleable output**
Sea buckthorn	up to 6m (20 ft)	low	edible berry
Hippophae rhamnoides	coppiceable nitrogen-fixing salt tolerant	April to May	berry leaf and bark: for culinary, cosmetic and medicinal uses woodchip (beware of thorns when handling)
Autumn olive, Russian olive	up to 3m (10 ft)	high	small, red edible berry
Elaeagnus umbellate, E. angustifolia	coppiceable nitrogen-fixing forage	June	none

Biodiversity benefits / Risks	How to fit into agroforestry system
Blossoms feed insects. Fallen fruit feeds birds and small mammals. susceptible to several pests and diseases because so widely cultivated	Like apples, the trees can be trained to almost any size and shape. When grown on vigorous rootstock, it can provide large and long-lived trees in hedgerows, or wider farms.
Flowers relatively early and feeds a range of insects. Fallen fruit feeds birds and small mammals. suckers prolifically (most species) cankers and silver leaf disease – increased risk from pruning in winter	Works great as part of single or mixed-fruiting row of medium-sized trees.
Early apricot blossoms feed bees and other insects. sucker prolifically (most species) peach leaf curl disease in damp climates needs good rich soil	In the UK, it might be best grown under cover or in a drier, warmer part of the country. Otherwise, it fits well in growing systems.

Biodiversity benefits / Risks	How to fit into agroforestry system
Dense, thorny branches provide nesting for birds. Fruit stays late on the bush, providing food for a range of wildlife. suckering potentially invasivea	Works as a good pioneer species, especially when grown in rows and the suckering is contained via cultivation or mowing. Works as a good nurse crop in windy situations. Roots are good for stabilising soil and preventing erosion.
Thorny branches provide nesting for birds. Berries feed a range of birds and mammals. potentially invasive, through self-seeding and spreading root systems	Works as a good infill plant among slower-growing fruit species Helps other trees establish.[2] Roots are good for stabilising soil and preventing erosion.

NUTS

Common name(s)	Typical height	Pollinator rating	Fruit/nut
Latin name	**Useful attributes**	**Flowering months**	**Other saleable output**
English walnut, heartnut	up to 35m (115 ft)	high	nuts
Juglans regia, J. ailantifolia	coppiceable	May to June	firewood timber woodchip
Sweet chestnut	up to 35m	high	nuts
Castanea sativa	coppiceable	July	fencing firewood timber woodchip
Almond	4–12m (13–39 ft)	high	nuts
Prunus amygdalus	potentially coppiceable	Feb to March	none

EVERGREEN

Common name(s)	Typical height	Pollinator rating	Fruit/nut
Latin name	**Useful attributes**	**Flowering months**	**Other saleable output**
Scots pine, black pine, stone pine	25–55m (82–180 ft), depending on species	low	seeds, more potential as climate warms
Pinus sylvestris, P. nigra, P. pinea	good in low-nutrient and drier soils	none	timber woodchip
Eucalyptus	10m–60m (33–197 ft), depending on species	high	none
Eucalyptus spp	coppiceable	May to August, depending on species	firewood foliage: floristry sales leaves: medicinal use timber woodchip

Biodiversity benefits	Risks	How to fit into agroforestry system
Many species of moth caterpillar eat the leaves. Nuts feed mammals, including mice and squirrels.	squirrels allelopathic effect from roots, inhibits growth of other plants nearby	Coppice for timber and shelter. Grow as standards for nut crop, but not too close to vegetable crops.
Good for red squirrels.	diseases such as blight and *Phytophthora* spp	Grow on long coppice rotation to provide managed shelter and saleable timber products.
One of the very first trees to blossom, provides early pollen and nectar for insects.	peach leaf curl susceptible to spring frosts (flowers so early) suckers freely	With upright growth and fine leaves, it is well suited to agro-forestry with other crops grown underneath.

Biodiversity benefits	Risks	How to fit into agroforestry system
Evergreen foliage provides winter cover for a range of species. In UK, it is an important habitat for red squirrels, pine martens and wildcats.	tendency for suppression of growth under the canopy	Use in shelterbelts and trees as hedges to provide winter protection. But they are too tall and shading for cropping areas.
Flowers for a long period in summer, so fills a useful gap for pollinators.	high water use suppression of growth under canopy	Repels flies, so good for systems with livestock. Grows quickly, so useful nurse crop for slow-growing species.

EVERGREEN

Common name(s)	Typical height	Pollinator rating	Fruit/nut
Latin name	**Useful attributes**	**Flowering months**	**Other saleable output**
Common holly	15m (49 ft)	high	none
Ilex aquifolium	coppiceable	May	firewood boughs with berries: Christmas market timber woodchip
Oaks (live, evergreen)	up to 40m (131 ft)	low	acorns: for flour
Quercus spp	coppiceable	May to June	firewood timber woodchip
Strawberry tree	up to 15m	medium	small red fruit
Arbutus unedo	salt tolerant	October to November	honey, potentially (in warmer climates) leaves: medicinal use
Bay tree	up to 15m	medium	none
Laurus nobilis	coppiceable	May to June	leaves: culinary herb and floristry
Manuka	4m (13 ft)	high	none
Leptospermum scoparium	salt tolerant	June	honey medicinal products (Yields should improve as climate warms.)

Biodiversity benefits / Risks	How to fit into agroforestry system
Provides cover and nesting for birds. Leaves are eaten by caterpillars of the holly blue butterfly and various moths. Berries are vital food for birds and small mammals in winter. — very tasty to deer relatively slow to establish	As a good evergreen choice, provides some winter protection in mixed rows.
Provides winter cover for a range of species If coppiced, it provides useful nesting sites for birds and mammals. Pollen and honeydew feed insects. — might suffer from oak decline possibly invasive in some areas	When coppiced within a mixed row, provides winter shelter.
Flowers are a significant source of nectar and pollen for insects. Fruit feeds birds. Evergreen foliage is good for nesting birds and winter cover. — possible damage at temperatures below –5°C (23°F)	Works well as medium-sized tree in mixed row.
Dense evergreen foliage is great nesting and cover for birds. — might not survive temperatures below –5°C (23°F) Grows poorly in wet, heavy soil	Small coppiceable evergreen species provides winter shelter within cropping area.
In its native New Zealand, provides a wide range of biodiversity benefits. Great for pollinators everywhere but wider benefits have not been widely studied. — Grows poorly in full shade	In the right conditions, it establishes quickly. Works best planted with other, smaller trees in low-growing alleys. Works well for coastal sites.

EVERGREEN

Common name(s)	Typical height	Pollinator rating	Fruit/nut
Latin name	**Useful attributes**	**Flowering months**	**Other saleable output**
Ivies	climbing species: up to 10m (33 ft) long	high	none
Hedera spp	forage	September to December	none
Ulmo	10m	high	none
Eucryphia cordifolia	coppiceable salt tolerant	July to August	honey firewood, in small quantities timber

SMALL

Common name(s)	Typical height	Pollinator rating	Fruit/nut
Latin name	**Useful attributes**	**Flowering months**	**Other saleable output**
Elder	6m (20 ft)	low	fruit: jams and wine
Sambucus nigra	none	June	flowers: a range of culinary and medicinal uses; dyes
Spindle	9m (30 ft)	medium	none
Euonymus europaeus	coppiceable	May to June	charcoal firewood oil: soap making timber: small specialist market, including artist woodchip

Biodiversity benefits	Risks	How to fit into agroforestry system
They are the last important nectar and honey plants of the season. They feed pollinating insects, including solitary bees. They may help some insect species establish a second generation in mild winters. Provides forage for butterflies and moths, is a caterpillar food plant, and provides berries in the winter for birds.	irritation from leaves and hairy stems	Interplant with vigorous tree species. Despite the myth, it will not kill or strangle a healthy tree.
Pollen and nectar are great for insects. Evergreen foliage provides cover and nesting for birds.	might not survive temperatures below −10°C (14°F)	Works well for coastal sites.

Biodiversity benefits	Risks	How to fit into agroforestry system
Provides food for a range of moth larvae. Nectar feeds a wide variety of insects. Fruit is eaten by birds and mammals, such as rodent species.	toxic: branches, twigs, leaves, roots and seeds (glycoside, can cause adverse reaction)	Great all-round small tree for mixed rows. Copes with dry soils.
Seeds feed later-season birds. Moth caterpillars eat leaves. Attracts both aphids and their predators. Nectar and pollen feed insects.	toxic: leaves and berries	Grows well in semi-shade. Works well planted between larger trees.

SMALL

Common name(s)	Typical height	Pollinator rating	Fruit/nut
Latin name	**Useful attributes**	**Flowering months**	**Other saleable output**
Common hawthorn	up to 15m (49 ft)	high	berries: culinary and medicinal uses
Crataegus monogyna	coppiceable	May	woodchip
Dogwoods	10m (33 ft)	high	none
Cornus spp	coppiceable	April to May	stems: coloured for winter floristry timber: specialist markets woodchip
Common buckthorn	8m (26 ft)	medium	none
Rhamnus cathartica	coppiceable not nitrogen-fixing but can boost soil nitrogen levels	April to May	woodchip
Guelder rose	4m (13 ft)	medium	berry: poisonous when raw, edible when cooked
Viburnum opulus	coppiceable	May to July	berry: medicinal market

Biodiversity benefits	Risks	How to fit into agroforestry system
Supports hundreds of species. Dormice eat the flowers. Nectar and pollen feed insects. Fruit is rich in antioxidants and is eaten by migrating birds and small mammals. Dense, thorny foliage provides nesting shelter for birds and food for caterpillars.	very thorny (as the name suggests) – wear gloves. susceptible to some apple diseases	Good in any mixed-tree row. Can help keep animals out of an area.
Moth caterpillars eat leaves. Bees, flies and beetles eat pollen and nectar. Birds and mammals eat berries.	no major risks	Coppice as low-level understorey for taller fruiting or timber trees. Cut every one or two years.
Good for pollinators. Food species for brimstone butterfly.	invasive species in North America, although can increase soil fertility[3]	Although not nitrogen-fixing, it can increase nitrogen and carbon levels in soil. It has potential use as fertility generator in cropped areas.
Supports a wide range of species, including moths and beetles. Berries feed birds and mammals.	host plant for black bean aphid	Works well as an interplant species amongst larger fruiting or nut trees.

MEDIUM

Common name(s)	Typical height	Pollinator rating	Fruit/nut
Latin name	**Useful attributes**	**Flowering months**	**Other saleable output**
Rowan, mountain ash	15m (49 ft)	medium	none
Sorbus aucuparia	coppiceable	May to June	firewood timber woodchip
Field maple	20m (66 ft)	high	none
Acer campestre	coppiceable	April to May	firewood timber sap: for syrup woodchip
Wild service tree	13–15m (43–49 ft)	Medium	fruit: historically used to make alcoholic drinks
Sorbus torminalis	coppiceable	May to June	firewood timber woodchip
Birches	30m (98 ft)	High	none
Betula spp	coppiceable	April to May	firewood sap: for processing timber woodchip

Biodiversity benefits / Risks	How to fit into agroforestry system
Leaves feed moth caterpillars. Flowers provide pollen and nectar for insects. Berries are a valuable food source in the autumn for birds like blackbirds and thrushes. *Risks:* toxic: raw berries	Grow as a medium tree or coppice within mixed/single row.
Attracts aphids and their predators, especially ladybirds and hoverflies. Leaves feed moths, including the mocha (*Cyclophora annularia*). Mammals eat fruits. *Risks:* no major risks	Grow as a standard tree or coppice within mixed/single row.
Leaves feed a range of moth and butterfly caterpillars. Blossoms give nectar and pollen to insects. Fruit feeds birds and small mammals. *Risks:* regular pruning if growing for timber to prevent forking	Grows well in an agroforestry setting because it needs good light to grow.
Light, open canopy is good for understorey plants. Provides food and habitat for over 300 insect species. Leaves feed many moth caterpillars. Associated with woodpeckers, fungi and hole-nesting birds. Seeds feed siskins, greenfinches and redpolls. *Risks:* very extensive, shallow rooting system	This quick-growing pioneer species helps other trees to establish. Light foliage is good for growing perennial shrubs and plants underneath in mixed-production row.

LARGE

Common name(s)	Typical height	Pollinator rating	Fruit/nut
Latin name	**Useful attributes**	**Flowering months**	**Other saleable output**
Whitebeam	15m (49 ft)	high	none
Aria edulis (syn. *Sorbus aria*)	coppiceable grows well in poor soils	May	firewood timber woodchip
Small-leaved lime, large-leaved lime	20m (66 ft)	high	edible fruit, but marginal commercially
Tilia cordata, *T. platyphyllos*	coppiceable forage	June to July	firewood flowers, leaves: teas, medicinal markets timber woodchip
Elms	Up to 30m (98 ft)	medium	none
Ulmus spp	coppiceable forage	January to March Some species: late summer or autumn	firewood timber woodchip
Oaks (deciduous)	40m (131 ft)	low	acorn: flour or livestock feed
Quercus spp	coppiceable	April to May	firewood timber woodchip
Beeches	40m	medium	nuts: edible but not widely marketed pannage for pigs
Fagus spp	coppiceable	April to May	firewood timber woodchip

Biodiversity benefits / Risks	How to fit into agroforestry system
Berries feed birds. Flowers feed bees and pollinating insects. Leaves are eaten by lots of moth caterpillars. — no known risks	Is useful as a nurse or pioneer species, especially in poorer and dry soils, which it will help improve.
These are big producers of nectar and pollen for insects. Also support lots of honeydew. Long-lived species provides habitat for birds and wood-boring beetles. — no known risks	Works well as a coppiced species within rows but will get big if left unmanaged. Alternatively, grow as standard away from cropped areas.
Very early flowering provides pollen for insects when not much else available. Seeds feed small mammals. — Dutch elm disease (though there are now resistant species on the market)	Works well as a coppiced species within rows but will get big if left unmanaged. Alternatively, grow as standard away from cropped areas.
Oak in the UK supports more life than any other tree species: hundreds of insects and fungi, as well as birds and mammals. — acute oak decline chronic oak decline oak processionary moth Suggest planting diverse genetics to build resilience.	Coppice or pollard on long cycles, or incorporate it as standard into non-cropped areas.
Foliage feeds moth caterpillars. Seeds feed mice, voles, squirrels and birds. Beech trees live long, so provide habitats for many deadwood specialists, like hole-nesting birds and wood-boring insects. — very susceptible to grey squirrel damage hard to establish well as timber tree	Coppice or pollard on long cycles, or incorporate it as standard into non-cropped areas.

LARGE

Common name(s)	Typical height	Pollinator rating	Fruit/nut
Latin name	**Useful attributes**	**Flowering months**	**Other saleable output**
Common hornbeam	30m (98 ft)	medium	none
Carpinus betulus	coppiceable	April to May	excellent firewood timber woodchip
Black locust	30m	high	seeds: edible to humans and livestock
Robinia pseudoacacia	coppiceable nitrogen-fixing	May to June	major honey plant fencing: very hard and rot-resistant firewood timber woodchip
Poplar, cottonwood, aspen	up to 50m (164 ft), depending on species	medium	none
Populus spp	coppiceable	February to March	biomass firewood: chipped timber woodchip
Sycamore	20–35m (66–115 ft)	high	none
Acer pseudo-platanus	coppiceable	April to May	firewood timber woodchip

Biodiversity benefits	Risks	How to fit into agroforestry system
Leaves feed a range of caterpillars. Pollen feeds insects. Seeds are particularly useful because they stay on the tree over winter, providing safe feeding for birds.	susceptible to *Phytophthora ramorum*	Coppice or pollard on long cycles, or incorporate it as standard into non-cropped areas.
Hosts more than 65 species of moth and butterfly.	suckers prolifically, possibly invasive	Due to suckering, it is not recommended in growing areas. Keep in rows or contained patches and away from vegetables, or in grazed part of the holding.
Provides food for moth caterpillars. Catkins provide early source of pollen and nectar for insects. Seeds feed birds.	dioecious: if seeds are the objective, grow female plants brittle wood, susceptible to storm damage: don't grow near tunnels or glasshouses	Coppice for high production on short rotation for woodchip.
This species is underrated for wildlife. Flowers feed insects. Seeds feed birds and small mammals. Provides habitat for birds, mammals, insects and fungi.	prolific self-seeder, invasive in some areas	Coppice on longer rotation to provide firewood, woodchip or timber.

RESOURCES

Books

Bukowski, Catherine and John Munsell, *The Community Food Forest Handbook: How to Plan, Organize, and Nurture Edible Gathering Places* (Chelsea Green Publishing, 2018).

Champagnat, Paul, *The Pruning of Fruit Trees*, tr. N. B. Bagenal (Crosby Lockwood and Son Ltd, 1954).

Crawford, Martin, *Shrubs for Gardens, Agroforestry and Permaculture* (Permanent Publications, 2020).

Crawford, Martin, *Trees for Gardens, Orchards and Permaculture* (Permanent Publications, 2015).

Foster, Tim, *Fruit for Life: A Friendly Guide to Growing Fruit Organically* (Eco-logic books, 2019).

Garrett, H. E., Shibu Jose, Michael A. Gold (eds), *North American Agroforestry* (John Wiley and Sons, 2022).

Gordon, Andrew and Steven Newman, *Temperate Agroforestry Systems* (CAB, 2006).

Goulson, Dave, *Silent Earth: Averting the Insect Apocalypse* (Vintage Publishing, 2021).

Hall, Jenny and Iain Tolhurst, *Growing Green: Animal-Free Organic Techniques* (Chelsea Green Publishing, 2010).

Krawczyk, Mark, *Coppice Agroforestry: Tending Trees for Product, Profit, and Woodland Ecology* (New Society Publishers, 2022).

Levy, Allyson and Scott Serrano, *Cold-Hardy Fruits and Nuts: 50 Easy-to-Grow Plants for the Organic Home Garden or Landscape* (Chelsea Green Publishing, 2022).

Nair, P. K. R., *An Introduction to Agroforestry* (Kluwer Academic Publishers, 1993).

O'Connell, Marina, *Designing Regenerative Food Systems: And Why We Need Them Now* (Hawthorn Press, 2022).

Phillips, Michael, *The Holistic Orchard: Tree Fruits and Berries the Biological Way* (Chelsea Green Publishing, 2012).

Phillips, Roger, Sheila Grant, and Tom Wellsted, *Trees in Britain, Europe and North America* (Pan Books, 1978).

Rackham, Oliver, *Ancient Woodland: Its History, Vegetation and Uses in England* (Edward Arnold, 1980).

Raskin, Ben and Simone Osborn (eds), *The Agroforestry Handbook: Agroforestry for the UK* (Soil Association, 2019).

Raskin, Ben, *The Woodchip Handbook: A Complete Guide for Farmers, Gardeners and Landscapers* (Chelsea Green Publishing, 2021).

Sheldrake, Merlin, *Entangled Life: How Fungi Make Our Worlds, Change Our Minds, and Shape Our Futures* (Vintage Publishing, 2023)

Stobart, Anne, *The Medicinal Forest Garden Handbook: Growing, Harvesting and Using Healing Trees and Shrubs in a Temperate Climate* (Permanent Publications, 2020).

Sumption, Phil, The *Organic Vegetable Grower: A Practical Guide to Growing for the Market* (Crowood Press, 2023).

Ward, Chloe, *How to Prune an Apple Tree: A Guide for Real People with Imperfect Trees* (IngramSpark, 2016).

Warlop, Francois, Nathalie Corroyer, Alice Denis et al., *Combining Vegetables and Fruit Trees in Agroforestry: Principles, Technical Elements and Points of Vigilance to Design and Lead His Plot (*Project Smart, 2017).

Weiseman, Wayne, Daniel Halsey, and Bryce Ruddock, *Integrated Forest Gardening: The Complete Guide to Polycultures and Plant Guilds in Permaculture Systems* (Chelsea Green Publishing, 2014).

Organisations and Websites

AFINET (Agroforestry Innovation Networks), agroforestrynet.eu/afinet

Agforward, www.agforward.eu

Agricology, agricology.co.uk

Agroforestry Research Trust, agroforestry.co.uk

Dartington Trust, 'Agroforestry at Dartington', dartington.org/about/our-land/agroforestry

Farm Woodland Forum, agroforestry.ac.uk: UK-based agroforestry organisation

Missouri Agricultural Experiment Station, University of Missouri, 'Horticulture and Agroforestry Research Farm', moaes.missouri.edu/central-missouri-research-extension-and-education-center/horticulture-and-agroforestry-research-farm

Orange Pippin Trees, orangepippintrees.co.uk: fantastic resource on a range of fruit trees

Organic Growers Alliance, organicgrowersalliance.co.uk

Organic Research Centre, organicresearchcentre.com

Orto Forest, www.ortoforesta.it: a living soil garden and regenerative market garden in Italy

Six Inches of Soil (film), sixinchesofsoil.org

Symphony of the Soil (film), symphonyofthesoil.com

Wakelyns, wakelyns.co.uk

NOTES

Introduction

1. M. E. Kislev, A. Hartmann and O. Bar-Yosef, 'Early Domesticated Fig in the Jordan Valley', *Science* 312/5778 (June 2006):1372–74, https://doi.org/10.1126/science.1125910.
2. G. Shepard, C. Clement, H. Lima, G. Mendes dos Santos, C. Moraes and E. Neves, 'Ancient and Traditional Agriculture in South America: Tropical Lowlands', *Oxford Research Encyclopedia: Environmental Science* (February 2020), https://doi.org/10.1093/acrefore/9780199389414.013.597.
3. A. Heinimann, O. Mertz, S. Frolking et al., 'A Global View of Shifting Cultivation: Recent, Current, and Future Extent', *PLoS ONE* 12/9: e0184479, https://doi.org/10.1371/journal.pone.0184479.

Chapter 1. Principles and Benefits of Silvohorticulture

1. Dev Gurera and Bharat Bhushan, 'Passive Water Harvesting by Desert Plants and Animals: Lessons from Nature', *Philosophical Transactions of the Royal Society*, https://doi.org/10.1098/rsta.2019.0444.
2. David E. Lott and Vaughn E. Hammond, *Water Wise Vegetable and Fruit Production*, University of Nebraska, Lincoln Extension, Institute of Agriculture and Natural Resources, NebGuide G2189.
3. Charles S. Baldwin, '10. The Influence of Field Windbreaks on Vegetable and Specialty Crops', *Agriculture, Ecosystems & Environment* 22–23 (1988): 191–203, https://doi.org/10.1016/0167-8809(88)90018-7.
4. Georgina Hennessy, Ciaran Harris, Charlotte Eaton et al., 'Gone with the Wind: Effects of Wind on Honey Bee Visit Rate and Foraging Behaviour', *Animal Behaviour* 161 (2020): 23–31, doi: 10.1016/j.anbehav.2019.12.018.
5. Laurie Hodges and Jim Brandle, 'Windbreaks for Fruit and Vegetable Crops', University of Nebraska-Lincoln, Extension Guide EC1779.
6. Laurie Hodges, Mohd Nazip Suratman, James R. Brandle and Kenneth G. Hubbard, 'Growth and Yield of Snap Beans as Affected by Wind Protection and Microclimate Changes Due to Shelterbelts and Planting Dates', *HortScience* 39/5 (2004): 996–1004, https://doi.org/10.21273/HORTSCI.39.5.996.
7. Timothy Hall, 'Paulownia: An Agroforestry Gem', *Trees for Life Journal* 3/3 (2008), https://tfljournal.org/article.php/20080418100402327.

8. Gordon Johnson, 'Expect Heat Damage to Vegetable and Fruit Crops', University of Delaware, *Weekly Crop Update* (22 July, 2022), https://sites.udel.edu/weeklycropupdate/?p=20781.
9. Agriculture at a Crossroads, Global Report (2009), International Assessment of Agricultural Knowledge, Science and Technology for Development.

Chapter 2. Site and Soil: Assessments and Implications

1. You can find this tool at https://www.forestresearch.gov.uk/tools-and-resources/fthr/ecological-site-classification/ (last access 23 July, 2024).
2. Z. Liu, W. Zhan, B. Bechtel et al., 'Surface Warming in Global Cities Is Substantially More Rapid Than in Rural Background Areas', *Communications Earth & Environment* 3/219 (2022), https://doi.org/10.1038/s43247-022-00539-x.
3. Ming Kuo and William Sullivan, 'Environment and Crime in the Inner City: Does Vegetation Reduce Crime?' *Acoustics, Speech, and Signal Processing Newsletter* 33/3 (May 2001): 343–67, https://doi.org/10.1177/00139160121973025.

Chapter 3. How Trees Interact with Crops and Each Other

1. M. R. Hoosbeek, R. P. Remme and G. M. Rusch, 'Trees Enhance Soil Carbon Sequestration and Nutrient Cycling in a Silvopastoral System in South-Western Nicaragua', *Agroforestery Systems* 92 (2018): 263–73, https://doi.org/10.1007/s10457-016-0049-2; P. Pardon, B. Reubens, D. Reheul et al., 'Trees Increase Soil Organic Carbon and Nutrient Availability in Temperate Agroforestry Systems', *Agriculture, Ecosystems & Environment* 247 (2017): 98–111.
2. N. R. Chari and B. N. Taylor, 'Soil Organic Matter Formation and Loss Are Mediated by Root Exudates in a Temperate Forest', *National Geoscience* 15 (2022): 1011–16.
3. Heidi-Jayne Hawkins, Rachael I.M. Cargill, Michael E. Van Nuland et al., 'Mycorrhizal Mycelium as a Global Carbon Pool', *Current Biology* 33/11 (2023): R560–73.
4. Mario Fontana, Alice Johannes, Claudio Zaccone et al., 'Improving Crop Nutrition, Soil Carbon Storage and Soil Physical Fertility using Ramial Wood Chips', *Environmental Technology & Innovation* 31 (2023): 103143, https://doi.org/10.1016/j.eti.2023.103143.
5. S. C. Wagner, 'Biological Nitrogen Fixation', *Nature Education Knowledge* 3/10 (20111):15.
6. Craswell Bijay-Singh, 'Fertilisers and Nitrate Pollution of Surface and Groundwater: An Increasingly Pervasive Global Problem', *SN Applied Science* 3 (2021): 518, https://doi.org/10.1007/s42452-021-04521-8.
7. Stavros Veresoglou, E. Barto, George Menexes and Matthias Rillig, 'Fertilization Affects Severity of Disease Caused by Fungal Plant Pathogens', *Plant Pathology* 62/5: 961–69, https://doi.org/10.1111/ppa.12014.

8. Kunlong Hui, Beidou Xi, Wenbing Tan and Qidao Song, 'Long-Term Application of Nitrogen Fertilizer Alters the Properties of Dissolved Soil Organic Matter and Increases the Accumulation of Polycyclic Aromatic Hydrocarbons', *Environmental Research* 215/2 (2022): 114267, https://doi.org/10.1016/j.envres.2022.114267.
9. H. Xu, M. Detto, S. Fang et al., 'Soil Nitrogen Concentration Mediates the Relationship between Leguminous Trees and Neighbour Diversity in Tropical Forests', *Communications Biology* 3 (2020): 317, https://doi.org/10.1038/s42003-020-1041-y.
10. D. Z. Epihov, K. Saltonstall, S. Batterman et al., 'Legume–Microbiome Interactions Unlock Mineral Nutrients in Regrowing Tropical Forests', *Proceedings of the National Academy of Sciences, USA* 118/11 (2021): e2022241118, https://doi.org/10.1073/pnas.2022241118.
11. N. M. Neves, R. R. Paula, E. A. Araujo et al., 'Contribution of Legume and Non-legume Trees to Litter Dynamics and C-N-P Inputs in a Secondary Seasonally Dry Tropical Forest', *iForest: Biogeosciences and Forestry* 15 (2022): 8–15, https://doi.org/10.3832/ifor3442-014.
12. K. S. Gangwar, S. K. Sharma and O. K. Tomar, 'Alley Cropping of *Subabul* (*Leucaena leucocephala*) for Sustaining Higher Crop Productivity and Soil Fertility of Rice (*Oryza sativa*)–Wheat (*Triticum aestivum*) System in Semi-arid Conditions', *Indian Journal of Agronomy* 49/2 (2004): 84–88.
13. C. R. Ward, D. R. Chadwick and P. W. Hill, 'Potential to Improve Nitrogen Use Efficiency (NUE) by Use of Perennial Mobile Green Manures', *Nutrient Cycling in Agroecosystems* 125 (2023): 43–62, https://doi.org/10.1007/s10705-022-10253-x.
14. Vishnu K. Solanki, Vinita Parte, J.S. Ranawat and R. Bajpai, 'Improved Nutrient Cycling and Soil Productivity through Agroforestry', *International Journal of Current Microbiology and Applied Sciences,* Special Issue 11 (2020): 1724–29.
15. L. Beule and P. Karlovsky, 'Early Response of Soil Fungal Communities to the Conversion of Monoculture Cropland to a Temperate Agroforestry System', *PeerJ* 9 (2021): e12236, https://doi.org/10.7717/peerj.12236.
16. C. Dupraz, C. Blitz-Frayret, I. Lecomte, Q. Molto, F. Reyes and M. Gosme, 'Influence of Latitude on the Light Availability for Intercrops in an Agroforestry Alley-Cropping System', *Agroforestry Systems* 92/4 (2018):1019–33, https://doi.org/10.1007/s10457-018-0214-x.
17. P. Pardon, B. Reubens, J. Mertens et al., 'Effects of Temperate Agroforestry on Yield and Quality of Different Arable Intercrops', *Agricultural Systems* 166 (2018): 135–51, https://doi.org/10.1016/j.agsy.2018.08.008.
18. Jo Smith, Ambrogio Costanzo, Nick Fradgeley, Samantha Mullender, Martin Wolfe and João Palma, *Lessons Learnt: Silvoarable Agroforestry in the UK* (Part 1), Agroforestry for Europe (2017): https://agforward.eu/documents/LessonsLearnt/WP4_UK_Silvoarable_1_lessons_learnt.pdf.

19. Bijay Tamang, Michael Andreu, Melissa Friedman and Donald Rockwood, 'Windbreak Designs and Planting for Florida Agricultural Fields', askIFAS, EDIS (2009): https://doi.org/10.32473/edis-fr289-2009.
20. John Davis, Lindsay Whistance and David Lewis, 'The Role and Benefits of Shelterbelts on Farms: A Vision for the Adoption of "Optimal Shelterbelts"', *Royal Forestry Society* 117/2 (April 2023): 115–21.
21. J. Sorribas, S. González, A. Domínguez-Gento and R. Vercher, 'Abundance, Movements and Biodiversity of Flying Predatory Insects in Crop and Non-crop Agroecosystems', *Agronomy for Sustainable Development* 36 (2016): 34, https://doi.org/10.1007/s13593-016-0360-3.
22. Johnathan Riley, 'Thriving Insects, Thriving Farm', The Furrow (February 2022): https://thefurrow.co.uk/thriving-insects-thriving-farm.
23. With thanks to APlus Tree Care and Sustainability (aplustree.com) and Orchard of Flavors (www.orchardofflavours.com) websites for inspiration on tree root types.
24. Bid Webb, 'Investigating the Impact of Trees and Hedgerows on Landscape Hydrology', PhD Theses, Bangor University, Wales, 2021.

Chapter 4. Identifying Objectives and Limiting Factors

1. M. W. Cadotte, 'Experimental Evidence that Evolutionarily Diverse Assemblages Result in Higher Productivity', *Proceedings of the National Academy of Science, USA* 110/22: 8996–9000, https://doi.org/10.1073/pnas.1301685110.

Chapter 5. Trees and Nature

1. W. J. Lewis, J. C. Lenteren, S. C. Phatak and J. H. Tumlinson, 'A Total System Approach to Sustainable Pest Management', *Proceedings of National Academy of Science, USA* 94/23 (1997): 12243–48, https://doi.org/10.1073/pnas.94.23.12243.
2. M. Barzman, P. Bàrberi, A. N. E. Birch et al., 'Eight Principles of Integrated Pest Management', *Agronomy for Sustainable Development* 35 (2015): 1199–215, https://doi.org/10.1007/s13593-015-0327-9.
3. Discussion on farm size at The Farming Forum, www.thefarmingforum.co.uk.
4. A. van der Wal, W. de Boer, W. Smant et al., 'Initial Decay of Woody Fragments in Soil Is Influenced by Size, Vertical Position, Nitrogen Availability and Soil Origin', *Plant Soil* 301 (2007): 189–201, https://doi.org/1007/s11104-007-9437-8.
5. To watch examples of the defense fungi, go to Journey to the Microcosmos at https://www.youtube.com/@journeytomicro.
6. M. S. Gorzelak, A. K. Asay, B. J. Pickles, S. W. Simard, 'Inter-plant Communication through Mycorrhizal Networks Mediates Complex Adaptive Behaviour in Plant Communities', *AoB Plants* 7 (May 2015): plv050, https://doi.org/10.1093/aobpla/plv050.
7. Department for Environment Food & Rural Affairs, 'Bees and Other Pollinators: Their Value and Health in England', July 2013, https://assets.publishing.service.gov.uk/media/5a7c620ae5274a7ee2567180/pb13981-bees-pollinators-review.pdf.

8. Keng-Lou James Hung, Jennifer M. Kingston, Matthias Albrecht, David A. Holway and Joshua R. Kohn, 'The Worldwide Importance of Honey Bees as Pollinators in Natural Habitats', *Proceedings of the Royal Society B* 285/1870 (2018): 20172140, https://doi.org/10.1098/rspb.2017.2140.
9. J. Bishop, M. P. D. Garratt and T. D. Breeze, 'Yield Benefits of Additional Pollination to Faba Bean Vary with Cultivar, Scale, Yield Parameter and Experimental Method', *Scientific Reports* 10 (2020): 2012, https://doi.org/10.1038/s41598-020-58518-1.
10. Angus Dingley, Sidra Anwar, Paul Kristiansen et al., 'Precision Pollination Strategies for Advancing Horticultural Tomato Crop Production', *Agronomy* 12/2: 518, https://doi.org/10.3390/agronomy12020518.
11. Jessica L Knapp and Juliet L Osborne, 'Courgette Production: Pollination Demand, Supply, and Value', *Journal of Economic Entomology* 110/5 (October 2017): 1973–79, https://doi.org/10.1093/jee/tox184.
12. J. G. Boyles, P. M. Cryan, G. F. McCracken and T. H. Kunz, 'Economic Importance of Bats in Agriculture', *Science* 332/6025: 41–2, https://doi.org/10.1126/science.120136.
13. Simon J. Grove, 'Saproxylic Insect Ecology and the Sustainable Management of Forests', *Annual Review of Ecology and Systematics* 33 (2002): 1–23, http://www.jstor.org/stable/3069254.
14. N. E. Stork, J. McBroom, C. Gely and A. J. Hamilton, 'New Approaches Narrow Global Species Estimates for Beetles, Insects, and Terrestrial Arthropods', *Proceedings of the National Academy of Sciences, USA* 112/24 (2015): 7519–23, https://doi.org/10.1073/pnas.1502408112.
15. B. Carbonne, S. Petit, V. Neidel et al. 'The Resilience of Weed Seedbank Regulation by Carabid Beetles, at Continental Scales, to Alternative Prey', *PNAS* 10 (2020): 19315, https://doi.org/10.1073/pnas.1502408112.
16. M. T. Fountain, 'Impacts of Wildflower Interventions on Beneficial Insects in Fruit Crops: A Review', *Insects* 13/3 (March 2022): 304. https://doi.org/10.3390/insects13030304; R. Long, A. Corbett, C. Lamb, C. Reberg-Horton, J. Chandler, M. Stimmann, 'Beneficial Insects Move from Flowering Plants to Nearby Crops', *California Agriculture* 52/5 (1998): 23–26, https://doi.org/10.3733/ca.v052n05p23.

Chapter 6. Getting Value from Your Trees

1. Royal Forestry Society, *The Cost of Grey Squirrel Damage to Trees in England and Wales* (January 2021): https://rfs.org.uk/wp-content/uploads/2021/03/grey-squirrel-impact-report-overview.pdf.
2. David Ibarra, Raquel Martín-Sampedro, Bernd Wicklein, Úrsula Fillat and María E. Eugenio, 'Production of Microfibrillated Cellulose from Fast-Growing Poplar and Olive Tree Pruning by Physical Pretreatment', *Applied Sciences* 11/14 (2021): 6445, https://doi.org/10.3390/app11146445.

3. Q. Fu, Y. Chen, M. Sorieul, 'Wood-Based Flexible Electronics', *ACS Nano* 14/3 (March 2020): 3528–38, https://doi.org/10.1021/acsnano.9b09817.

Chapter 7. Designing the Layout

1. Leon Bessert, 'Keyline Design – Water Management of Agricultural Landscapes: Key for Regenerative Agriculture?', University of Kassel (March 2022): https://agroforst-info.de/wp-content/uploads/2023/01/bessert-Keyline_angepasst.pdf.
2. Tom Staton, Richard Walters, Jo Smith, Tom Breeze and Robbie Girling, 'Management to Promote Flowering Understoreys Benefits Natural Enemy Diversity, Aphid Suppression and Income in an Agroforestry System', *Agronomy* 11/4 (2021): 651, https://doi.org/10.3390/agronomy11040651.
3. Tellie-Nelson Wajja-Musukwe, Julia Wilson, Janet I. Sprent, Chin K. Ong, J. Douglas Deans, John Okorio, 'Tree Growth and Management in Ugandan Agroforestry Systems: Effects of Root Pruning on Tree Growth and Crop Yield', *Tree Physiology* 28/2 (February 2008): 233–42, https://doi.org/10.1093/treephys/28.2.233; G. R. Korwar and G. D. Radder, 'Influence of Root Pruning and Cutting Interval of *Leucaena* Hedgerows on Performance of Alley Cropping *Rabi* Sorghum', *Agroforestry Systems* 25 (1994): 95–109, https://doi.org/10.1007/BF00705670.

Chapter 8. Tree Planting and Establishment

1. Gabriel Schenker, Armando Lenz, Christian Körne and Günter Hoch, 'Physiological Minimum Temperatures for Root Growth in Seven Common European Broad-leaved Tree Species', *Tree Physiology* 34/3 (March 2014): 302–13, https://doi.org/10.1093/treephys/tpu003.
2. Ben Raskin, *The Woodchip Handbook: A Complete Guide for Farmers, Gardeners, and Landscapers* (White River Junction, VT: Chelsea Green Publishing, 2021).
3. David Granatstein and Kent Mullinix, 'Mulching Options for Northwest Organic and Conventional Orchards', *HortSci* 43/1 (2008): 45–50, https://doi.org/10.21273/HORTSCI.43.1.45.
4. David Granatstein and Kent Mullinix, 'Mulching Options for Northwest Organic and Conventional Orchards', *HortSci* 43/1 (2008): 45–50, https://doi.org/10.21273/HORTSCI.43.1.45.
5. H. An, F. Luo, T. Wu et al., 'Dwarfing Effect of Apple Rootstocks Is Intimately Associated with Low Number of Fine Roots', *HortSci* 52/4 (2017): 503–12, https://doi.org/10.21273/HORTSCI11579-16.

Chapter 9. Early-Years Management

1. 'Investigating the Efficacy of Willow Woodchip Mulch as a Control for Apple Scab', Soil Association, https://www.innovativefarmers.org/field-labs/investigating-the-efficacy-of-willow-woodchip-mulch-as-a-control-for-apple-scab.

Appendix. Tree Directory

1. Broads Authority, 'Riverbank Protection Works: A Guide for Riparian Landowners', Broads Design & Management Information (2004): https://www.broads-authority.gov.uk/__data/assets/pdf_file/0033/182697/Riverbank_Protection_Works.pdf.
2. Jo Clark and Gabriel Hemery, 'The Use of the Autumn Olive in British Forestry', *Quarterly Journal of Forestry* 100 (2006): 285–88, https://www.futuretrees.org/wp-content/uploads/2020/10/Clark-and-Hemery-2006-Use-of-autumn-olive-in-british-forestry.pdf.
3. Liam Heneghan, Farrah Fatemi, Lauren Umek, Kevin Grady, Kristen Fagen and Margaret Workman, 'The Invasive Shrub European Buckthorn (*Rhamnus cathartica*, L.) Alters Soil Properties in Midwestern US Woodlands', *Applied Soil Ecology* 32/1 (May 2006): 142–48, https://doi.org/10.1016/j.apsoil.2005.03.009.

INDEX

S

ABOUT THE AUTHORS

Andy Dibben

Andy Dibben is currently Head Grower at Abbey Home Farm, where he is responsible for the production of over ninety different fruit and vegetable crops. Over the last eight years, Andy has been responsible for designing and establishing an agroforestry system integrating fruit, coppice and wildlife trees with field-scale vegetable production. Andy regularly writes horticultural articles and speaks at conferences, including the Oxford Real Farming Conference.

Ivan Raskin

Ben Raskin has worked in the commercial growing world for more than thirty years and has a wide range of practical horticultural experience, having started his career growing vegetables for chefs, box schemes and farm shops. He now focuses mainly on integrating trees into farming systems. Ben is head of agroforestry for the Soil Association and keeps his hands in the field implementing a pioneering agroforestry system on a 1,500-acre farm in Wiltshire in South West England. He also works as an independent advisor, trainer and author, writing books for children and grownups, and is chair of the Community Supported Agriculture Network UK.

the politics and practice of sustainable living

CHELSEA GREEN PUBLISHING

Chelsea Green Publishing sees books as tools for effecting cultural change and seeks to empower citizens to participate in reclaiming our global commons and become its impassioned stewards. If you enjoyed *Silvohorticulture*, please consider these other great books related to sustainable farming and agroforestry.

THE WOODCHIP HANDBOOK
A Complete Guide for Farmers, Gardeners and Landscapers
BEN RASKIN
9781645020486
Paperback

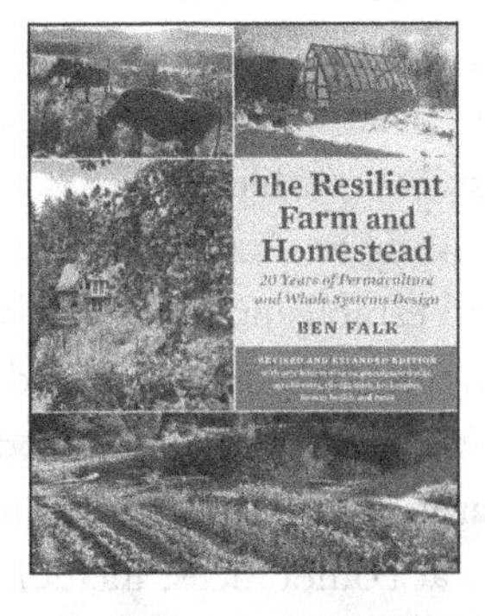

THE RESILIENT FARM AND HOMESTEAD, REVISED AND EXPANDED EDITION
20 Years of Permaculture and Whole Systems Design
BEN FALK
9781645021100
Paperback

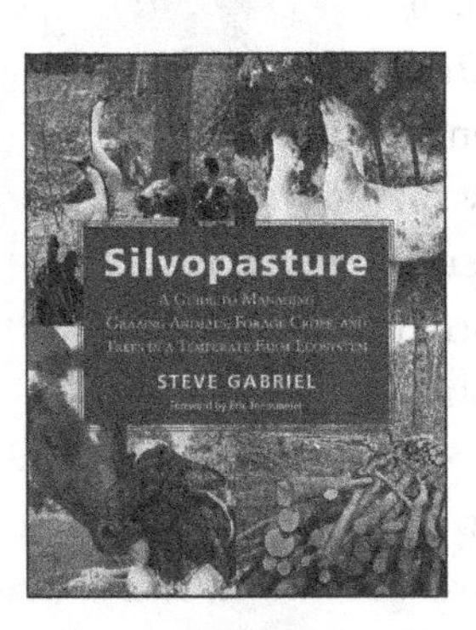

SILVOPASTURE
A Guide to Managing Pasture Animals, Forage Crops, and Trees in a Temperate Farm Ecosystem
STEVE GABRIEL
9781603587310
Paperback

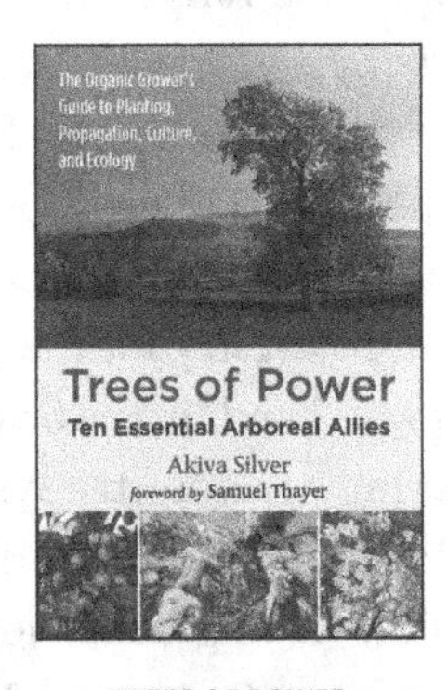

TREES OF POWER
Ten Essential Arboreal Allies
AKIVA SILVER
9781603588416
Paperback

For more information,
visit **www.chelseagreen.com.**